RESIDUE REVIEWS

VOLUME 37

Phosphamidon

SINGLE PESTICIDE VOLUME:

PHOSPHAMIDON

RESIDUE REVIEWS

Residues of Pesticides and Other
Foreign Chemicals in Foods and Feeds

RÜCKSTANDS-BERICHTE

Rückstände von Pestiziden und anderen
Fremdstoffen in Nahrungs- und Futtermitteln

Editor
FRANCIS A. GUNTHER

Assistant Editor
JANE DAVIES GUNTHER

Riverside, California

VOLUME 37

SPRINGER-VERLAG
BERLIN · HEIDELBERG · NEW YORK
1971

ISBN 978-1-4615-8475-9   ISBN 978-1-4615-8473-5 (eBook)
DOI 10.1007/978-1-4615-8473-5

Preface

More and more biologists, chemists, pharmacologists, toxicologists, governmental agencies, and "food control" (regulatory) officials around the world are finding it increasingly difficult to keep abreast of the technical literature in the pesticide field; indeed, many libraries do not have even a small proportion of the journals and other sources that now regularly contain research, development, and application information about all aspects of modern chemical pest control. As a result, a very large number of requests has come to "Residue Reviews" to publish detailed digests of information on single pesticide chemicals so that the interested person in any part of the world could easily be brought up to date with all available important information without having to search probably several hundred literature sources, many of them obscure or simply not available except in very large libraries. The service and convenience rendered the readership by such a series of volumes on major individual pesticide chemicals would therefore be considerable.

Type and scope of coverage in this series of single-pesticide volumes will of course vary with available information. The coverage should be as complete as possible, however, to be of maximum value to all interested individuals, industries, research institutions, and governmental agencies concerned with the continuing production of an adequately large yet safe food supply for the world. Among the topics bracketed for a single pesticide should ideally be:

I. Introduction
II. History of development and use, including alternate names around the world, patent information
III. Chemistry, manufacture, and stability
    a) Synthesis
    b) Chemical and physical properties
    c) Commercial synthesis and composition of commercial product
    d) Formulations
    e) Storage stability of formulated products
    f) Compatability with other materials
    g) Photodecomposition
    h) Stability of parent compound in water, animal and plant tissues, and soils
    i) Other known chemical and biochemical reactions
IV. Pharmacology and toxicology
    a) Acute, chronic, dermal and inhalation toxicities

    b) Metabolic pathways and products in animals, plants, and other soils
    c) Other important effects on mammals
    d) Effects on wildlife in controlled tests

  V. Biological properties and uses, world coverage

    a) Pests and crops involved
    b) Performance
    c) Methods of application
    d) Dosages and formulations
    e) Any secondary effects (e. g., phytotoxicity with insecticides or fungicides)

  VI. Tolerances, world coverage

  VII. Residues encountered in foodstuffs in developmental programs and in practice, with dosages, timing, methods of application, persistence curves, half-lives, etc.

    a) Raw agricultural commodities
    b) Processed food products
    c) Residue removal by washing, processing, etc.
    d) Metabolites or other alteration products as residues

  VIII. Analytical methods, with full details of recommended methods

    a) Assay of technical grade product
    b) Assay of formulated products
    c) Residue methods for foodstuffs, waters, soils, including recoveries, minimum detectabilities, sensitivities, precision, etc. both in the presence of and in the absence of substrate extractives

  IX. Incidence in, and effects on, environment

    a) Incidence in the environment
    b) Persistence in the environment
    c) Soil-compound interactions (soil flora and fauna)
    d) Stability in soils, water, and soil water
    e) Effects on fish and other wildlife, pollinating insects
    f) Involvement in the food chain

  X. Discussion and conclusions

Individual chapters or sections may be credited to particular persons as authors, of course, but it is suggested that a single person be responsible for tone, style, and acceptable "Residue Reviews" format of the total manuscript.

As with other volumes of "Residue Reviews", manuscripts are normally contributed by invitation, but in English only. Preliminary communication with the editor is necessary before volunteered reviews are submitted in manuscript form.

Department of Entomology   Francis A. Gunther
University of California   Jane Davies Gunther
Riverside, California
February 17, 1971

Foreword

In 1956 *CIBA Ltd.,* Basle, Switzerland, at that time a newcomer in the field of pesticide research and development, reported on a new organophosphorus insecticide at the annual meeting of the "Centre International des Antiparasitaires" in Lugano. The compound carried the code number C-570, and was a vinyl-type phosphate. It represented a chemical group in which *CIBA Ltd.* had shown particular interest since 1951, the year of the company's first vinylphosphate patent on C-177, dichlorvos. Compound C-570, which is marketed under the trade name DIMECRON®, was given the common name "phosphamidon", this name being recognized by both the British Standards Institution and the American Standards Organization. Phosphamidon was the first well-known representative from among the group of *N*-alkylated carbamyl vinylphosphates, to which group *Shell* and *CIBA* have more recently added two further insecticides, dicrotophos (C-709; BIDRIN®, CARBICRON®) and monocrotophos (C-1414; AZO-DRIN®, NUVACRON®).

Phosphamidon has been and still is one of the major agrochemical proucts of *CIBA Ltd.* However, its use in large-scale control operations is changing from the classical high volume application to ultra-low volume, waterless spraying by aircrafts. *CIBA Ltd.'s* recent rice project contracted with the Indonesian Government is a typical example of modern and efficient insect control in the world's most important crop.

The worldwide use of phosphamidon against sucking insects on a variety of agricultural commodities has laid the foundations for a steady growth of *CIBA's* Agrochemical Division. Today the company is not only marketing vinyl-type phosphorus esters, but also insecticidal carbamates, and the new formamidine-type acaricide, chlorphenamidine. Several substituted ureas and a diphenylether compound have opened the way into the vast herbicide market. Some of the insecticides are not only used in the field of plant protection, but have shown promise for the control of arthropods in stored products as well as in animal and public hygiene.

The present volume on phosphamidon attempts to summarize and evaluate the numerous field and laboratory experiments carried out by the company and many institutes and cooperators during the past decade. As stressed by the editor of "Residue Reviews", special emphasis has been given to those topics that are related to public health and safety-in-use aspects of the compound. The different review articles have been prepared by *CIBA's* specialists, who are aware of the fact that their papers reflect an immense amount of hard and often troublesome research and development work conducted by many colleagues and technicians throughout the world. Of particular value were the many significant contributions made by the scientists of the former *California Chemical Company*, now *Chevron Chemical Company*, whose experiments brought to light important features of the insecticide. These data and results were vital for registration of phosphamidon in the United States and elsewhere.

CIBA Ltd. expresses its gratitude to all these many unnamed people for their tireless efforts and devotes to them the present comprehensive account on phosphamidon.

For the Management and
Scientific Staff of *CIBA Ltd.'s*
Agrochemical Division

Basle (Switzerland) 1971   Henri Martin
                                            Josef Schuler

Table of Contents

SINGLE PESTICIDE VOLUME:

PHOSPHAMIDON

Chapter 1

Chemical and physical properties of phosphamidon

By

R. ANLIKER and E. BERIGER

Contents

I. Introduction

Phosphamidon ($C_{10}H_{19}ClNO_5P$) is a broad-spectrum systemic insecticide (BACHMANN *et al.* 1956 and 1957) of the vinyl phosphate group, comprising a mixture of *cis-* and *trans*-isomers in the proportion 73 : 27. It was first prepared by BERIGER and SALLMANN *(CIBA Ltd.* 1955, ANLIKER *et al.* 1961 a).

The importance of the product necessitated studies of its hydrolysis under various conditions, and synthesis and identification of its possible breakdown products. The experimental findings are reviewed in the present chapter, and the assignment of the configuration of the α- and β-isomers of phosphamidon and its derivatives is discussed.

For investigations on the metabolism of organophosphorus insecticides labeling with ^{32}P was preferred because of the high specific activity and ready availability of this tracer. However, this method provides only limited information on the metabolic fate of the non-phosphorus fractions of the insecticides. To obtain more complete information, it was necessary

1

to use phosphamidon labeled at various sites on the molecule. Metabolic studies with phosphamidon labeled at more than one site are described in Chapter 4.

II. Chemistry of phosphamidon

The following section covers the chemistry of phosphamidon, *i.e.*, the chemical names of the compound, laboratory and technical syntheses, and the chemical properties.

a) Chemical names

The various chemical names of phosphamidon are listed below:
2-chloro-N,N-diethyl-3-hydroxycrotonamide, dimethylphosphate (name in
    accordance with rules of nomenclature used in *Chemical Abstracts*)
O,O-dimethyl-O-[2-chloro-2-(diethylcarbamoyl)-1-methylvinyl]-
    phosphate
dimethylphosphate of 2-chloro-3-hydroxy-N,N-diethylcrotonamide
O-(1-chloro-1-diethylcarbamoyl-1-propen-2-yl)-O,O-dimethylphosphate
O,O-dimethyl-O-(diethylamido-1-chloro-crotonyl-2)-phosphate
2-(O,O-dimethylphosphoryloxy)-1-chloro-crotonic acid diethylamide
2-chloro-3-(dimethoxyphosphinyloxy)-N,N-diethylcrotonamide

b) Laboratory synthesis

Phosphamidon is prepared by reaction of trimethylphosphite with α,α-dichloroacetoacetic acid diethylamide, which is obtained by chlorination of acetoacetic acid diethylamide. The reaction proceeds rapidly in boiling chlorobenzene, yielding a constant mixture of *cis*- and *trans*-isomers in the approximate proportion 73 : 27. This ratio remains practically unchanged when the conditions of the reaction are varied (e. g., when catalysts of isomerisation such as acetic acid are added) or when the reaction mixture is exposed to ultraviolet light or heat. This situation contrasts with that found with, for example, dicrotophos [3-(dimethoxy-phosphinyloxy)-N,N-dimethylcrotonamide] (Menzer and Casida 1965). The major isomer, β-phosphamidon, has the *cis*-configuration and is considerably more active as a cholinesterase inhibitor and more toxic to mammals and insects than the *trans*-isomer, α-phosphamidon.

$$(CH_3O)_3P + CH_3COCCl_2CON(C_2H_5)_2$$

$trans(\alpha)$-Phosphamidon (27%) $cis(\beta)$-Phosphamidon (73%)
low biological activity   high biological activity

Radioactively labeled phosphamidon was synthesised for metabolic studies (ANLIKER et al. 1961 b) (see chapter 4). The synthesis used yielded phosphamidon of high specific activity, with a double label of ^{14}C:

$$2\ CH_3{}^{14}CON(C_2H_5)_2 \xrightarrow{POCl_3} CH_3{}^{14}COCH_2{}^{14}CON(C_2H_5)_2$$

$$\xrightarrow{SO_2Cl_2} CH_3{}^{14}COCCl_2{}^{14}CON(C_2H_5)_2 \xrightarrow{(CH_3O)_3P}$$

$$(CH_3O)_2P\overset{\displaystyle O}{\underset{\displaystyle O{}^{14}C}{\Big\langle}} \overset{CH_3}{\underset{}{\underset{}{|}}} \overset{Cl}{\underset{}{|}}$$

$$O{}^{14}C{=\!=\!=}C{-}{}^{14}CON(C_2H_5)_2$$

^{14}C-labeled acetic acid diethylamide (MURRAY-WILLIAMS 1958) was condensed (ANLIKER et al. 1961 b) by a modification of the method of BREDERECK et al. (1959) with phosphorus oxychloride to the double-labeled acetoacetic acid diethylamide. Chlorination by sulfuryl chloride yielded the α,α-dichloroacetoacetic acid diethylamide, which was condensed with trimethylphosphite to the ^{14}C-double-labeled phosphamidon.

This type of labeling permits the fate of even the smallest parts of the molecule to be followed. BULL et al. (1967) carried out their metabolic studies with phosphamidon labeled with ^{32}P and ^{14}C. CLEMONS and MENZER (1968) used phosphamidon-N,N-α-diethylamide-^{14}C, prepared by reacting N,N-diethylacetoacetamide-α-diethylamide-^{14}C with sulfuryl chloride and trimethylphosphite, using the method of ANLIKER et al. (1961 a).

c) Technical synthesis

Technical phosphamidon, a mixture of the two geometrical isomers, contains two major by-products having the vinyl phosphate structure, namely dechlorophosphamidon VIII [3-(dimethoxyphosphinyloxy)-N,N-diethylcrotonamide] and γ-chlorophosphamidon IX [2,4-dichloro-3-(dimethoxyphosphinyloxy)-N,N-diethylcrotonamide] (see Table V). Both compounds are mixtures of cis- and trans-isomers. During the chlorination of acetoacetic acid diethylamide to α,α-dichloroacetoacetic acid diethylamide, only small quantities of the α-monochlorinated and the α,α,γ-trichlorinated acetoacetic acid diethylamide are formed. These products are not removed during the process of manufacture, and they react with trimethylphosphite to form compounds VIII and IX.

Gas chromatographic analyses of the product of normal technical synthesis show the presence of four compounds, listed in Table I.

Table I. Composition of technical phosphamidon

Total vinylphosphates	92–94 %
trans-phosphamidon	24.8%
cis-phosphamidon	67.2%
dechlorophosphamidon (cis- and trans-)	~1 %
γ-chlorophosphamidon (cis- and trans-)	~1 %

The technical product is a brownish yellow, slightly aromatic liquid, sold under the trade marks DIMECRON ® and Famfos ®.

d) Chemical properties

Phosphamidon is stable at room temperature, but starts to decompose slowly at temperatures above 160° C. The addition of bromine to the double bond occurs at temperatures as low as 25° C.

Phosphamidon is most stable in mildly acid aqueous solution. The vinyl phosphate group is the most readily hydrolysed.

Table II. *Stability of phosphamidon in aqueous solutions*

Temperature (°C.)	Half-lives		
	pH 4	pH 7	pH 10
23	74 days	13.8 days	2.2 days
45	6.6 days	2.1 days	3.3 hours

Table II gives the half-life values as a function of pH *(California Chemical Company 1961)*. The products of breakdown are dimethyl phosphate and α-chloroacetoacetic acid diethylamide (Table V, compound X). In an analogous manner the principal metabolite, *N*-desethylphosphamidon (Table V, compound III), breaks down to α-chloroacetoacetic acid ethylamide (Table V, compound XI). The hydrolytic cleavage of the P-OCH$_3$ bond is very slow. This cleavage to O-desmethylphosphamidon (Table V, compound VI) also occurs *in vivo*, and at a measurable rate. This compound can be prepared by anionic demethylation with sodium iodide, or by reaction of phosphamidon with dodecylmercaptan and tetramethyl-ammonium hydroxide. It can be quantitatively converted back to phosphamidon by esterification with diazomethane, and this reaction can be used to detect O-desmethylphosphamidon (Table V, compound VI) in plant material *(CIBA Ltd. 1964 a)*. Even under weakly alkaline conditions the products of hydrolysis (Table V, compounds X and XI) are rapidly transformed to the hydroxylated products diethylamide and monoethylamide (Table V, compounds XII and XIII). This reaction is the basis of the positive and very sensitive reaction of phosphamidon and its chlorinated derivates (Table V) with tetrazolium blue, which is typical of acetoin structures. It is used for the detection of phosphamidon and some of its metabolites on paper chromatograms *(CIBA 1964 b, Pack et al. 1964)*. In alkaline solution acetoin (Table V, compound XII) rearranges to glycolic acid diethylamide (Table V, compound XIV) and acetic acid. The key step in this transformation is formally reversal of a Claison condensation. The degradation of phosphamidon follows a different course under strongly acid conditions. There is simultaneous hydrolysis of the enol phosphate and of the acid amide, with formation of the intermediate α-chloroacetoacetic acid, which is then decarboxylated to chloroacetone.

III. Physical properties of phosphamidon

In the following section a summary of the physical properties of phosphamidon, its by-products, and its metabolite N-desethylphosphamidon will be given. The assignment of a configuration on the basis of infrared (IR) spectra and biological tests will be discussed.

Table III. *Physical properties of phosphamidon*

Empirical formula	$C_{10}H_{19}NO_5ClP$
Molecular weight	299.5
Cis-trans ratio	73 : 27 (approx.)
Boiling points	150° C. at 1 mm. Hg
	115° C. at 0.2 mm. Hg
	94° C. at 0.04 mm. Hg
Specific gravity d_4^{25}	1.2132
Refractive index n_D^{25}	1.4721
Vapour pressures	$2.5 \cdot 10^{-5}$ mm. Hg at 20° C.
	$8.4 \cdot 10^{-5}$ mm. Hg at 30° C.
	$2.6 \cdot 10^{-4}$ mm. Hg at 40° C.
	$9.12 \cdot 10^{-4}$ mm. Hg at 60° C.
Volatility	0.41 mg./m.³ at 20° C.
	1.33 mg./m.³ at 30° C.
	4.00 mg./m.³ at 40° C.
Partition coefficients at 25° C.	hexane/water 0.083
	methylene chloride/water 90
Physical state	colorless liquid
Odor	faint, pleasant
Solubility	miscible with water and all common organic solvents; slightly soluble in saturated hydrocarbons (3.2% in hexane at 25° C.)

Table IV. *Physical properties of by-products of phosphamidon*

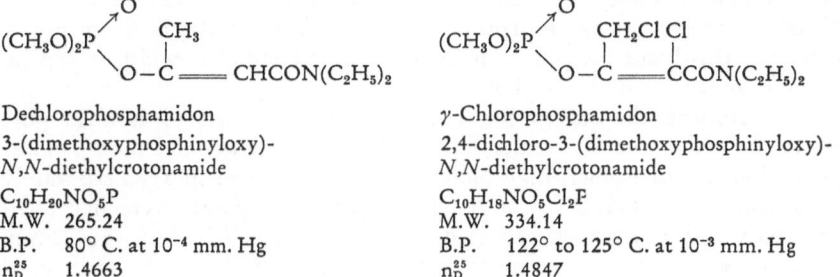

Dechlorophosphamidon
3-(dimethoxyphosphinyloxy)-
N,N-diethylcrotonamide
$C_{10}H_{20}NO_5P$
M.W. 265.24
B.P. 80° C. at 10^{-4} mm. Hg
n_D^{25} 1.4663

γ-Chlorophosphamidon
2,4-dichloro-3-(dimethoxyphosphinyloxy)-
N,N-diethylcrotonamide
$C_{10}H_{18}NO_5Cl_2P$
M.W. 334.14
B.P. 122° to 125° C. at 10^{-3} mm. Hg
n_D^{25} 1.4847

a) General physical properties

Tables III and IV summarise the typical physical properties of phosphamidon and its by-products.

b) Chromatographic behaviour

Table V summarises some data on the chromatographic behaviour of phosphamidon and its metabolites, degradation products, and by-products. All compounds having an α-hydroxy keto structure, or which yield such structures under alkaline conditions, give a blue colour reaction with an alkaline solution of blue tetrazolium chloride. The intensity of the colour reaction varies from compound to compound.

c) cis-trans Isomerism

Phosphamidon consists of a mixture of *cis-trans* isomers, which were separated by OSPENSON and PACK (1962) using gas, paper, and thin-layer chromatography. The assignment of a configuration based on NMR spectra is not conclusive. Studies by ANLIKER (1963) indicate that *cis*-phosphamidon (in which the allylic methyl and carbamoyldiethylamide groups are *cis* to each other) is the more active isomer as in the case of dicrotophos, mevinphos (FUKUTO *et al.* 1961, STOTHERS and SPENCER 1961) and monocrotophos (MENZER and CASIDA 1965). The assignment of configuration was made on the basis of IR spectra and the direct interrelation between the separated isomers of phosphamidon and its principal metabolite, *N*-desethylphosphamidon. This metabolite is synthesised by reacting trimethylphosphite with α,α-dichloroacetoacetic acid ethylamide. The isomers are formed in an approximate 67:33 *cis*-to-*trans* ratio, and can be readily separated by countercurrent distribution *(CIBA Ltd.* 1964 d).

In the IR spectrum the minor isomer, α-*N*-desethylphosphamidon, shows an absorption at 3350 cm.$^{-1}$, which does not disappear after strong dilution (0.004 molar). If this absorption band at 3350 cm.$^{-1}$ can be associated with intramolecular hydrogen bonding, this isomer is assigned the *trans*-structure. Intramolecular hydrogen bonding can only occur in the *trans* form. Then the major and biologically more active isomer, β-*N*-desethylphosphamidon, has the *cis*-structure. In concentrated solutions the *cis*-isomer shows a weak absorption band at 3350 cm.$^{-1}$ also in the infrared; however, this band disappears in highly diluted samples, which is typical for intermolecular hydrogen bonding.

To establish the correlation between the two pairs of isomers pure α-phosphamidon (minor isomer, low biological activity) and pure β-phosphamidon (major isomer, high biological activity) were sprayed separately onto bean plants *(Vicia faba)*. After 72 hours, the plants were harvested and analysed by the paper chromatographic method *(CIBA Ltd.* 1964, PACK *et al.* 1964). In the experiment with α-phosphamidon only unchanged

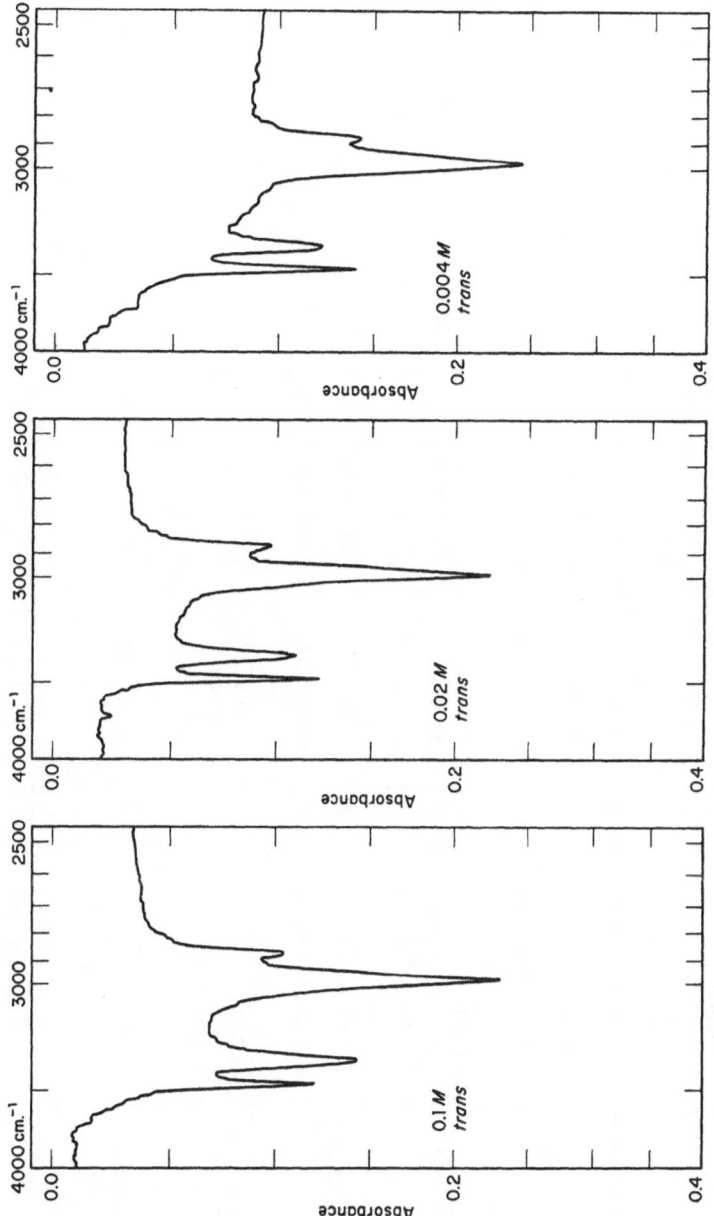

Fig. 1. Infrared spectra of *trans-N*-desethylphosphamidon in carbon tetrachloride

Table V. *Chromatographic behaviour of phosphamidon and its metabolites, products of decomposition, and by-products*

Compound no.	Product	M.P. (°C.) and B.P. (°C./mm. Hg)	Paper and thin-layer chromatography Solvent systems [a] and R_f values					Reaction with tetrazolium blue	Detected as metabolite
			A	B	C	D (cm.)	E		
I	$(CH_3O)_2P(O)$—O—C=C(CH_3Cl)·$CON(C_2H_5)_2$ [d] *cis*	130–135/0.001	0.40	0.94	1.00	7.00	0.51	pos.	—
	trans	130–135/0.001	0.40	0.94	1.00	11.00	0.51		
II	$(CH_3O)_2P(O)$—O—C=C(CH_3Cl)·CON$\langle \frac{C_2H_4OH}{C_2H_5}$ [c, e]	not isolated	—	0.75	—	—	—	pos.	yes
III	$(CH_3O)_2P(O)$—O—C=C(CH_3Cl)·$CONHC_2H_5$ [d] *cis*	38–40	0.23	0.88	1.00	2.00	—	pos.	yes
	trans	130–135/0.001	0.14	0.88	1.00	2.00	—		
IV	$(CH_3O)_2P(O)$—O=C—C(CH_3Cl)·$CONH_2$ [c, d]	oil, dec. on dist.	0	—	—	—	—	pos.	no
V	$(CH_3O)_2P(O)$—O—C=C(CH_3Cl)·COOH [c, d]	oil, 118–120 [b]	0	0.41	0.55	0.00	(F) 0.53	pos.	no
VI	$\frac{CH_3O}{HO}P(O)$—O—C=C(CH_3Cl)·$CON(C_2H_5)_2$ [c, d]	oil, dec. on dist.	0	0.55	0.67	0.00	(F) 0.75	pos.	yes

			A						
VII	$\begin{array}{l}CH_3O\\\nearrow^{O}\\P\\HO\end{array}\quad \begin{array}{l}CH_3Cl\\|\\O-C=C\cdot CONHC_2H_5\end{array}$ [c,e]	not isolated	0	0.55	0.67	0.00	—	pos.	no
VIII	$(CH_3O)_2P\nearrow^{O}\quad \begin{array}{l}CH_3\\|\\O-C=CHCON(C_2H_5)_2\end{array}$ [c,f]	80/10⁻⁴	0.35	—	—	—	—	neg.	by-product of tech. phosphamidon
IX	$(CH_3O)_2\overset{O}{P}\quad \begin{array}{l}CH_2Cl\ Cl\\|\\O-C=C\cdot CON(C_2H_5)_2\end{array}$ [c,f]	122–125/10⁻³	0.67	—	—	—	0.66	pos.	by-product of tech. phosphamidon
X	$CH_3COCHClCON(C_2H_5)_2$ [d]	78–80/0.01	0.69	1.00	1.00	26.00	—	pos.	yes
XI	$CH_3COCHClCONHC_2H_5$ [d]	44.5–45.5 / 91–92.5/0.1	0.53	1.00	1.00	19.00	0.68	pos.	yes
XII	$CH_3COCHOHCON(C_2H_5)_2$ [d,g]	73–75/0.45	0.38	—	—	—	—	pos.	no
XIII	$CH_3COCHOHCONHC_2H_5$ [h]	oil, dec. on dist.	0.46	—	—	—	0.49	pos.	no
XIV	$HOCH_2CON(C_2H_5)_2$ [d]	37–40/0.05	—	—	—	—	0.37	neg.	no
XV	$HOCH_2CONHC_2H_5$ [f]	102/0.01	—	—	—	—	1.00	neg.	no
XVI	CH_3COCH_2Cl [d]	119/736	—	—	—	—	—	pos.	no

[a] Solvent systems:

A Descending, Whatman No. 1 paper, pure gasoline b. r. 90°–95° C. (or heptane)-toluene-methanol-water (10:10:14:6 v/v) upper phase (ANLIKER et al. 1961).

B Ascending, Whatman No. 3 paper, acetonitrile-water-ammonium hydroxide (40:9:1, v/v) (BULL et al. 1967).

C Thin-layer, microcristalline cellulose, same solvent mixture as system B (ibid.).

D Descending, Whatman No. 3 paper impregnated in DMF (20 percent in acetone), cyclohexane saturated with dimethylformamide. Mobility expressed as cm. a compound migrated in 15 hours (ibid.).

E Thin-layer, silica gel G, ether-acetonitrile (3:1, v/v) (PACK et al., 1964).

F Descending, Whatman No. 1 paper, n-propanol-ethylacetate-water (7:1:2, v/v) (ANLIKER et al. 1961).

[b] m-Xylylthiuronium salt.　[d] ANLIKER et al. (1961).　[f] BERIGER E., private communication.　[h] Synthesised according to SZE (1965).

[c] cis-trans Mixtures.　[e] BULL et al. (1967).　[g] Industrial Bio-Test Laboratories Inc. (1965).

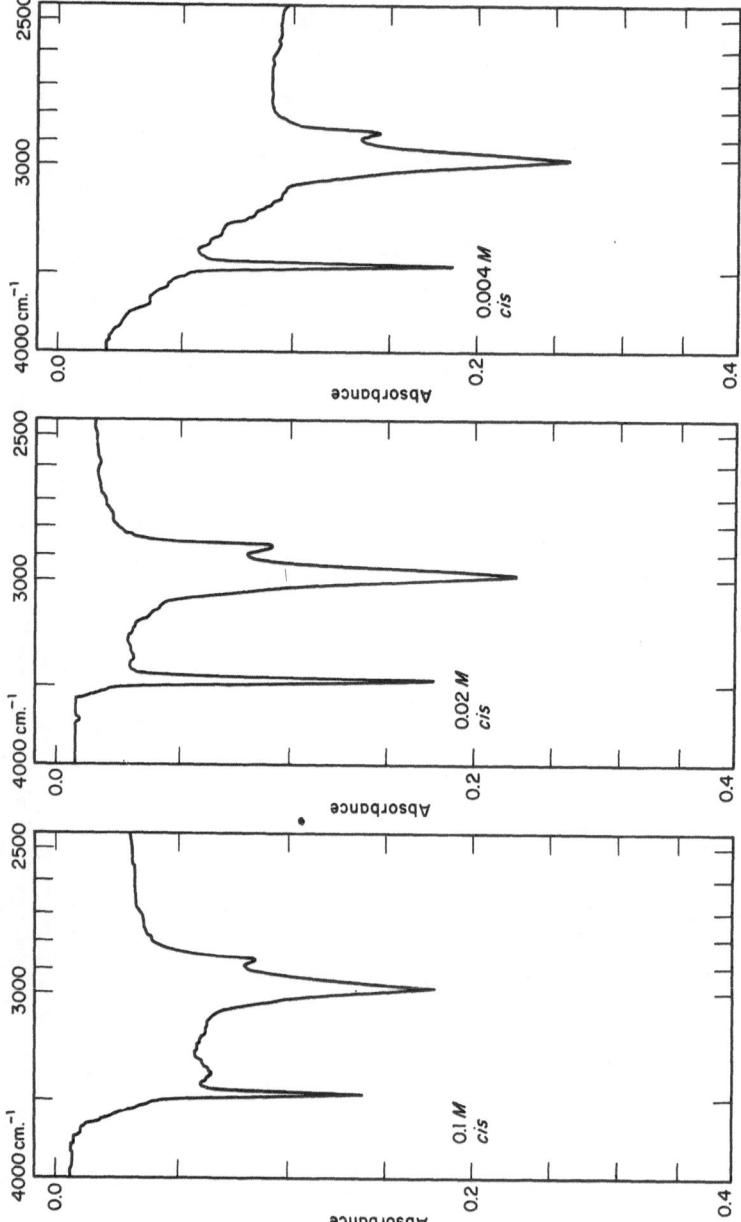

Fig. 2. Infrared spectra of *cis-N*-desethylphosphamidon in carbon tetrachloride

α-phosphamidon and *trans*-N-desethylphosphamidon, and with β-phosphamidon only unchanged β-phosphamidon and *cis*-N-desethylphosphamidon were isolated.

Since in the living organism no isomerisation of phosphamidon isomers has been observed so far, and there is little probability of an inversion during the enzymatic degradation of α- and β-phosphamidon to the corresponding N-desethylphosphamidon isomers under the given conditions, it may be concluded that the two α-isomers on the one hand and the two β-isomers on the other hand have the same configuration. As *cis*-N-desethylphosphamidon (β) arises from β-phosphamidon, the latter must have the *cis*-structure, and so α-phosphamidon must have the *trans*-structure.

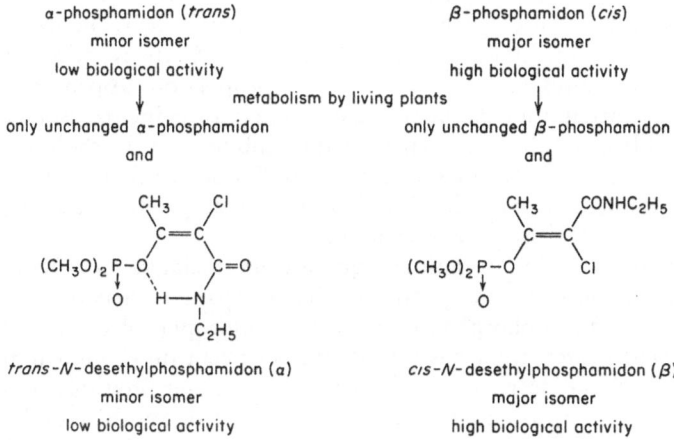

Fig. 3. Biological transformation of phosphamidon isomers to desethylphosphamidon isomers

Summary

Phosphamidon (2-chloro-N,N-diethyl-3-hydroxycrotonamide dimethylphosphate) is prepared by reacting trimethylphosphite with α,α-dichloroacetoacetic acid diethylamide. The reaction yields a constant mixture of *cis*- and *trans*-isomers in the approximate proportion 70:30. The assignment of the *cis*-configuration to the major isomer, which is considerably more active as a cholinesterase inhibitor, is supported by its infrared spectra and by the existence of a direct interrelation between the separated isomers of phosphamidon and those of its biological degradation product N-desethyl phosphamidon.

Technical phosphamidon, a mixture of the two mentioned isomers, contains two by-products having the vinyl phosphate structure, *i.e.*, dechlorophosphamidon and γ-chlorophosphamidon. Both of these compounds represent no more than one percent of the technical material.

In addition to describing the physical and chemical properties of phosphamidon, the present article summarizes methods for the synthesis of the isotopically labeled insecticide and of potential metabolites. Solvent systems and detecting reagents which were found to be useful for the chromatographic separation of phosphamidon, its by-products, and its metabolites on paper and thin-layer plates are listed.

Résumé *

Propriétés physiques et chimiques du phosphamidon

Le phosphamidon (2-chloro-N,N-diéthyl-3-hydroxycroton amide diméthylphosphate) est synthétisé en faisant réagir le triméthyl phosphite avec l'acide α,α-dichloroacétique diéthylamide. Cette réaction mène à un mélange d'isomères cis- et trans-, dont les proportions approximatives sont de 70/30. L'attribution de la configuration cis au principal isomère, qui est considérablement plus actif en tant qu'inhibiteur de la cholinestérase, est renforcée par le spectre infrarouge et par l'existence d'une relation directe entre les isomères séparés du phosphamidon et du produit de sa dégradation biologique: N-deséthylphosphamidon.

Le phosphamidon technique, qui est un mélange des deux isomères mentionnés, contient deux produits mineurs ayant la structure vinyl-phosphate: le deschlorophosphamidon et le γ-chlorophosphamidon. Ces deux composés ne représentent pas plus de un pour cent du produit technique.

Cet article ne décrit pas seulement les propriétés physiques et chimiques du phosphamidon. Il englobe aussi les méthodes de synthèse de l'insecticide marqué par radioisotope et des métabolites possibles. Dans un tableau sont réunis les systèmes de solvants ainsi que les réactifs qui permettent la révélation du phosphamidon et de ses produits de dégradation lors de leur séparation par chromatographie sur papier et sur couches minces.

Zusammenfassung **

Chemische und physikalische Eigenschaften des Phosphamidons

Phosphamidon, O,O-Dimethyl-O-[1-chlor-1-(N,N-diäthylcarbamyl)-propen-1-yl-2]-phosphat, wird durch Umsetzung von Trimethylphosphit mit Dichloracetessigsäurediäthylamid synthetisiert. Die Reaktion führt zu einem Gemisch von cis- und trans-Isomeren im ungefähren Verhältnis von 70 : 30. Die Annahme, daß die stärker cholinesterasehemmende Hauptisomere die cis-Konfiguration besitzt, erfolgte aufgrund von Infrarotspektren und vergleichenden Untersuchungen an den Isomeren des Phosphamidons und seines biologischen Abbauproduktes, N-Desäthylphosphamidon.

---

* Traduit par J. P. Lang.
** Übersetzt von den Autoren.

Technisches Phosphamidon besteht aus einer Mischung der beiden obengenannten Isomeren. Es enthält weiterhin zwei Nebenprodukte, die ebenfalls Vinylphosphat-Struktur besitzen: Deschlorphosphamidon und γ-Chlorphosphamidon. Die Konzentration beider Substanzen im technischen Material liegt im allgemeinen unter einem Prozent.

Der vorliegende Artikel beschreibt nicht nur die physikalischen und chemischen Eigenschaften des Phosphamidons. Er faßt auch die Methoden zur Synthese des radioaktiv markierten Insektizids sowie möglicher Metaboliten zusammen. Lösungsmittel-Systeme und Nachweisreagenzien, die bei papier- oder dünnschichtchromatographischer Trennung von Phosphamidon, seinen Nebenprodukten und Metaboliten von Nutzen sind, werden in einer Tabelle aufgeführt.

References

ANLIKER, R.: Metabolism of the systemic insecticide phosphamidon. Presented Vth Internat. Pesticide Congress, London (1963).
—, and R. E. MENZER: Method for phosphamidon residue analysis. J. Agr. Food Chem. 11, 391 (1963).
—, E. BERIGER, M. GEIGER, and K. SCHMID: Über die Synthese von Phosphamidon und seinen Abbau in Pflanzen. Helv. Chim. Acta 44, 1622 (1961 a).
— —, and K. SCHMID: Die Synthese von ¹⁴C-markiertem Phosphamidon, einem neuen systemischen Insektizid. Experientia 17, 492 (1961 b).
BACHMANN, F.: Phosphamidon, ein neuer Phosphorsäureester mit systemischer Wirkung. Proc. IVth Internat. Congress Crop Protection (Hamburg) 2, 1153 (1957).
—, and J. MEIERHANS: Un nouvel insecticide systémique, le Phosphamidon. Bull. Centre Internat. des Antiparasitaires, p. 18, Nov. (1956).
BREDERECK, H., R. GOMPPER, and K. KLEMM: Selbstkondensation N,N-disubstituierter Säureamide. Ber. 92, 1456 (1959).
BULL, D. L., D. A. LINDQUIST, and R. R. GRAPPE: Comparative fate of the geometric isomers of phosphamidon in plants and animals. J. Econ. Entomol. 60, 332 (1967).
California Chemical Co. (now Chevron Chemical Co.): Rate of hydrolysis of phosphamidon. Unpublished report (1961).
— Phosphamidon — preliminary report on the isolation and properties of α and β isomers. Unpublished report (1962).
— The appearance and decay of γ-chlorophosphamidon in crops sprayed with phosphamidon. Unpublished report (1964).
CIBA Ltd.: E. BERIGER, and R. SALLMANN, Swiss Pat. 342558, application 3. 11. 1955, patented 30. 11. 1959; U. S. Pat. 2,908,605.
— Metabolism of phosphamidon in plants. Identification of demethylphosphamidon in plant extracts. Unpublished report (1964 a).
— Analysis of phosphamidon residues. Paper chromatography method. Unpublished report (1964 b).
— Method for demethylphosphamidon residue analysis. Unpublished report (1964 c).
— Isolation and identification of α- and β-isomers of desethylphosphamidon. Unpublished report (1964 d).
CLEMONS, G. P., and R. E. MENZER: Oxidative metabolism of phosphamidon in rats and a goat. J. Agr. Food Chem. 16, 312 (1968).
FUKUTO, T. R., E. O. HORNING, R. L. METCALF, and M. Y. WINTON: Configuration of the α- and β-isomers of methyl 3-(dimethoxyphosphinyloxy)-crotonate (Phosdrin®). J. Org. Chem. 26, 4620 (1961).
Industrial Bio-Test Laboratories Inc.: Hydrolysis of N,N-diethyl-α-chloroacetoacetamide in mild alkaline medium. Unpublished report (1965).

Menzer, R. E., and J. E. Casida: Nature of toxic metabolites formed in mammals, insects, and plants from 3-(dimethoxyphosphinyloxy)-N,N-dimethyl-*cis*-crotonamide and its N-methyl analog. J. Agr. Food Chem. **13**, 102 (1965).

Murray, A., and D. L. Williams: Organic synthesis with isotopes, Part I. New York–London: Interscience (1958).

Pack, D. E., J. N. Ospenson, and G. K. Kohn: Phosphamidon. In G. Zweig, ed: Analytical methods for pesticides, plant growth regulators, and food additives, vol. II, p. 375. New York–London: Academic Press (1964).

Spencer, E. Y.: Nomenclature of phosphamidon isomers. J. Econ. Entomol. **60**, 1749 (1967).

Stothers, J. B., and E. Y. Spencer: Nuclear magnetic resonance spectra of a vinyl-phosphate and thionophosphate isomers (phosdrin and thionophosdrin). Can. J. Chem. **39**, 1389 (1961).

Chapter 2

Analytical methods for phosphamidon

By

WILLY BÜCHLER

Contents

I. Introduction

Phosphamidon, the dimethyl phosphate of *N,N*-diethyl-2-chloro-3-hydroxy crotonamide,

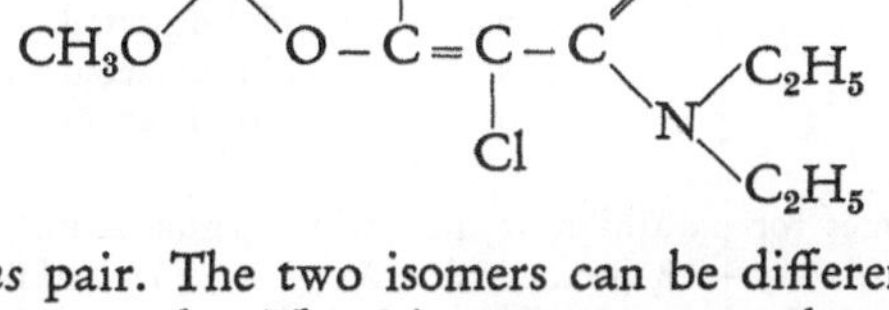

exists as a *cis-trans* pair. The two isomers can be differentiated by NMR or gas liquid chromatography. The *β*-isomer corresponds to the *cis*-form.

15

The over- or underchlorinated products γ-chlorophosphamidon and dechlorophosphamidon are present as impurities:

γ-chlorophosphamidon

dechlorophosphamidon

Table I. *Gas chromatographic retention times of organophosphorus pesticides*

Pesticide	Column temperature (°C.)	Relative retention time at the temperature shown [a]	
		SE-30 column	Apiezon column
Azinphos methyl	190	1190	1160
Demeton-O-methyl	No sample available	No sample available	No sample available
Demeton-S-methyl	165	33	21
Diazinon	165	38	32
Dichlorvos	125	4.5	2.6
Dimefox	125	2.9	1.6
Dimethoate	165	78	50
Disulfoton	165	41	42
Ethion	190	244	265
Formothion	165	91	67
Malathion	165	91	67
Mecarban	165	142	122
Menazon	190	Not detected	Not detected
Mevinphos	165	14	8.5
Morphothion	190	325	285
Oxydemeton-methyl	190	Not detected	Not detected
Parathion	165	100	100
Phenkapton	190	510	825
Phorate	165	25	25
Phosphamidon	165	1 peak 121	2 peaks 46/66
Schradan	165	Not detected	106
Trichlorphon	125	Not detected	Not detected
Vamidothion	190	Not detected	318

[a] The retention times for parathion at the various temperatures were SE-30 column = 190° — 3.25 min., 165° — 8.8 min., and 125° — 43 min.; Apiezon column = 190° — 3.25 min., 165° — 6.3 min., and 125° — 30 min.

They can be separated from phosphamidon by gas liquid chromato-graphy or thin-layer chromatography.

PACK *et al.* (1964) reviewed the available analytical methods, and since that time the iodometric method has been improved upon and several new methods have been developed.

II. Identification and detection

Phosphamidon can be detected by means of tetrazolium blue (MADER 1952), 4-(4-nitrobenzyl)-pyridine (GETZ and WATTS 1964), which reacts rapidly and with a high sensitivity, or by decomposition with acid and subsequent detection of phosphorus. These methods are, however, somewhat time-consuming.

J. RUZICKA *et al.* (1967) reported the gas chromatographic retention times of organophosphorus pesticides shown in Table I.

In some cases identification is possible from the IR-spectrum (CROSBY 1964, see below), and the NMR-spectrum also provides valuable infor-mation (see below).

The identification of phosphamidon in residues is described in the chapter on residue determination.

III. Quantitative determination

a) Iodometric method

Phosphamidon is reacted with iodine in strongly alkaline solution (ANLIKER *et al.* 1961), four equivalents of iodine being consumed. In sodium carbonate solution, however, only the by-products, mainly N,N-diethyl-2-chloroacetoacetamide and N,N-diethyl-2,2'-dichloroacetoacetamide, con-sume iodine, and thus the content of active substance may be determined by differential titration.

Procedure:

Accurately weigh out about 1.5 g. of the phosphamidon to be analysed, dissolve it in water, and dilute to 200.0 ml. Pipette 10.0 ml. of this solution into an Erlen-meyer flask fitted with a ground glass stopper and add 20.0 ml. of 0.1 N iodine solution, preferably by a piston burette, taking care to agitate the flask. Add 20 ml. of 2 N sodium hydroxide, mix the solution well, and let it stand in the dark for 30 minutes at $25 \pm 2°$ C. Then acidify with 20 ml. of 5 N hydrochloric acid and titrate the excess iodine with 0.1 N sodium thiosulfate, using starch as indicator.

Titre = u ml., so consumption of 0.1 N iodine solution = $(20 - u)$ ml. = a.

Pipette a further 10.0 ml. of the above phosphamidon solution into a second Erlenmeyer flask; while agitating the solution, add 5.0 ml. of 0.1 N iodine solution and 5 ml. of 2 N sodium carbonate. Stopper and allow to stand in the dark for 15 minutes at $25 \pm 2°$ C. Then acidify with 5 ml. of 5 N hydrochloric acid and titrate as above with 0.1 N thiosulfate.

Titre = v ml., so consumption of 0.1 N iodine solution = $(5 - v)$ ml. = b.

One ml. 0.1 N iodine corresponds to 7.493 mg. of phosphamidon.

Table II. *Survey of the methods for the quantitative determination of phosphamidon*

Method	Reaction or reacting group	Specificity	Application	Relative standard deviation ($s_{rel.}$, %)
Iodometric	Oxidation	Phosphamidon + γ-chloro-phosphamidon + dechloro-phosphamidon	Active subst. + formulations	1.0
Argentometric	Cl	Phosphamidon + γ-chloro-phosphamidon	Formulations	0.30
Mercaptometric	$-OCH_3$	Phosphoric ester	—	0.50
TLC-colorimetric	P	Phosphamidon	Stability studies	~1
TLC-colorimetric	$-N(C_2H_5)_2$	Phosphamidon	Stability studies	~2
Infrared	—	Phosphamidon or carbonyl group	Combinations	1–2
NMR	—	Phosphamidon	—	—
Gas chromatographic	—	Phosphamidon	Det. of impurities	0.4
Enzymatic	Cholinesterase inhibition	—	Automated determinations	1

Since the iodometric determination is an empirical method, the experimental conditions described above should be closely adhered to, particularly with regard to the suggested sequence of addition of iodine and sodium hydroxide. It is suitable for the determination of technical grade substances alone and in formulated preparations.

If by-products interfere with the determination of technical formulations, the active substance must be extracted and the organic solvents evaporated off before the determination is carried out.

b) Argentometric method

The total chlorine content is determined following alkaline hydrolysis. When the cleavage is carried out in piperidine (VOEGELI and CHRISTEN 1968), only the chlorine of monochloroacetic acid diethylamide and one of the two chlorine atoms of γ-chlorophosphamidon are released; these are determined separately. Dichloroacetic acid diethylamide is separated by partition between hexane and water.

Potassium hydroxide in propylene glycol: Dissolve 120 g. of chloride-free potassium hydroxide in 70 ml. of water and dilute to one litre.

Procedure:

Transfer 3.0 g. of phosphamidon, accurately weighed, into a separating funnel; add 45 ml. of water, 5 ml. of 2 N nitric acid, and 50 ml. of n-hexane, and shake well for one minute. Collect the clear aqueous phase in a 100-ml. volumetric flask.

Repeat the extraction of the hexane solution twice, first with 25 ml. and then with 20 ml. of water. Combine the aqueous extracts in the volumetric flask and dilute to the mark with water.

Reflux 10.0 ml. of this solution with 50 ml. of potassium hydroxide in propylene glycol for one hour. Cool, acidify with 80 ml. of 2 N nitric acid, and titrate potentiometrically with 0.1 N silver nitrate.

Consumption = a ml. 0.1 N AgNO$_3$/g. of phosphamidon.

Add, with agitation, a mixture of 5 ml. of water, 5 ml. of isopropanol, and 10 ml. of piperidine to 3.0 g. of phosphamidon, accurately weighed; cool (20° to 30° C.) and allow to stand for 30 minutes. Acidify with 80 ml. of 2 N nitric acid and titrate potentiometrically with 0.1 N silver nitrate.

Consumption = b ml. of 0.1 N AgNO$_3$/g. of phosphamidon.

The phosphamidon content is calculated as follows: Percent phosphamidon = 2.997 (a − b).

The second titration detects all the chlorine of any chloroacetic acid diethylamide present, as well as exactly one chlorine atom of γ-chlorophosphamidon.

With this method γ-chlorophosphamidon is expressed as phosphamidon, while dechlorophosphamidon is not detected. Dichloroacetic acid diethylamide remains quantitatively in the hexane phase during partition and is thus removed.

c) Mercaptometric method

Mercaptan reacts with phosphamidon in alkaline solution, resulting in cleavage of a methyl ester group. The excess mercaptan is back-titrated with standard iodine. Interfering impurities can be removed by extraction with bisulfite and petroleum spirit.

The mercaptan method detects not only phosphamidon but also γ-chlorophosphamidon and dechlorophosphamidon, all three being regarded as active compounds.

d) Colorimetric method

After separation by thin-layer chromatography, the silica gel zone containing the phosphamidon is removed from the plate. The phosphamidon is hydrolyzed by heating with sulfuric acid and the diethylamide moiety is reacted with cuprous diethyldithiocarbamate, the resulting complex being determined spectrophotometrically. The TLC separation renders the method specific, but does not yield absolute values.

Extraction solution: Mix 25 ml. of carbon disulfide with 475 ml. of chloroform.

Reagent solution: Add a solution of 155 g. of sodium acetate in 150 ml. of water to a solution of 5 g. of copper(II)-sulfate in 200 ml. of ammonia, 25 percent w/w.

Procedure:

After separation by thin-layer chromatography, the zone corresponding to phosphamidon is identified in UV light at 360 nm and removed quantitatively from the plate.

Place this silica gel powder in a 20-ml. stoppered test tube with a boiling stone, add 1.0 ml. of 50 percent sulfuric acid and allow the mixture to boil gently under reflux for one hour. When the mixture has cooled down, rinse the condenser with a small quantity of water.

Pour the acid hydrolysate thus obtained into a 250-ml. separating funnel and rinse the test tube with a small quantity of water. Add 2 ml. of conc. ammonia, 20 ml. of reagent solution, and 50 ml. of extraction solution and shake the mixture vigorously for one minute. When the lower phase has separated sufficiently, run it into a second separating funnel. Shake the aqueous solution once more with 20 ml. of extraction mixture. Transfer this extract into the second separating funnel and shake the combined extracts with 150 ml. of water. Filter the upper layer through a funnel containing a pledget of chloroform-washed cotton wool into a 100-ml. volumetric flask. Extract the aqueous phase with 20 ml. of chloroform and pass the extract into the volumetric flask. Dilute to the mark with chloroform (= test solution). A reference solution is obtained by treating 10.0 ml. of standard solution (0.720 g. of diethylamine in 100 ml. of water) with 20 ml. of reagent solution and extracting as mentioned above.

The absorbance of the test and reference solutions is measured against chloroform in 1-cm. cells at 435 nm.

e) Infrared method

The infrared spectrum of phosphamidon (Figure 1) has two bands which are suitable for quantitative determinations: the first at 1650 cm.$^{-1}$ in the spectra of chloroform solutions, arising from $C=O$ and $C=C$ stretching vibrations (LICHTENTHALER 1961), and the second at 860 cm.$^{-1}$ in chloroform solution spectra or at 890 cm.$^{-1}$ in the spectra of carbon tetrachloride solutions of the compound. By these methods, phosphamidon can often be specifically determined in admixture with other active substances.

The IR-spectra of α- and β-phosphamidon can be distinguished by the bands at 840 cm.$^{-1}$ and 950 cm.$^{-1}$, which occur only in the spectra of the α- and β-isomers, respectively, as well as those in the range from 1100 to 1300 cm.$^{-1}$

f) Nuclear magnetic resonance (NMR) method

The two isomers are easily distinguishable in the NMR spectrum. These bands are suitable for quantitative determinations (California Chemical Company):

Groups	Chemical shift (p.p.m)	
	α	β
CH$_3$-vinyl	2.38	2.10
CH$_3$O –	3.83	3.92
	4.0	4.1

g) Gas chromatographic method

Phosphamidon can be separated into its two isomers and quantitatively determined using the silicone oils SE-52 or SE-30 as stationary phases. Aluminum or glass columns may be used (Figure 2). The column tempera-

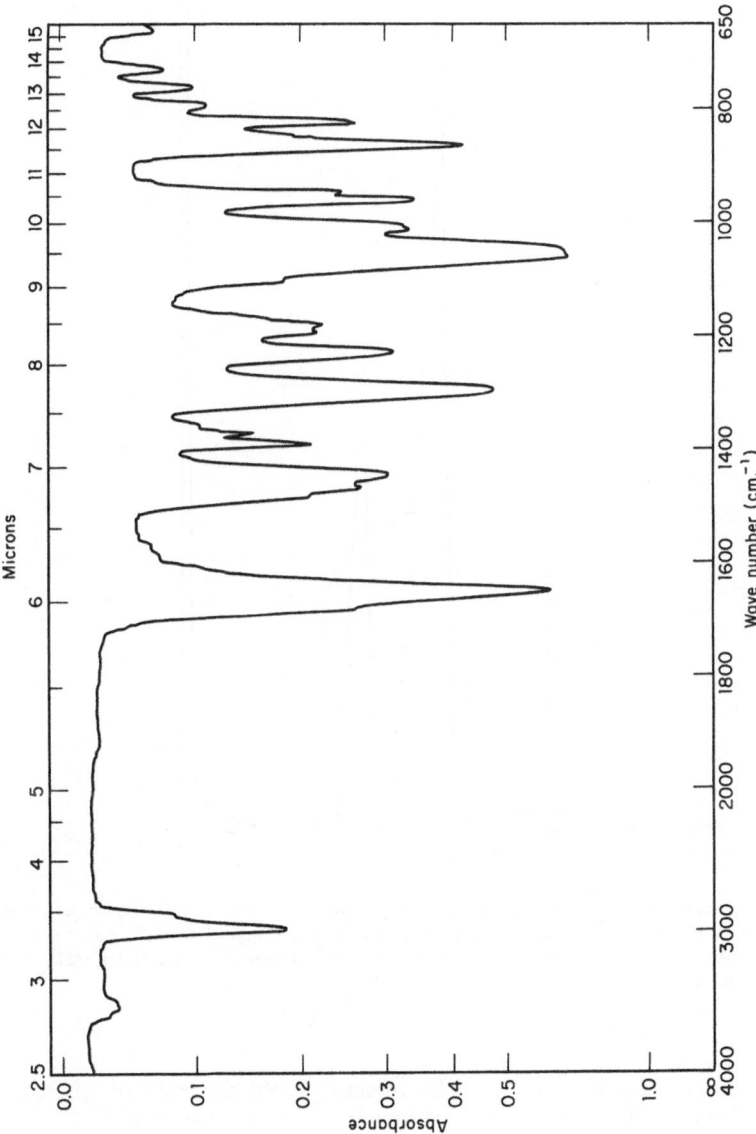

Fig. 1. Infrared spectrogram of phosphamidon: instrument, Perkin-Elmer 157; technique, capillary film between sodium chloride plates

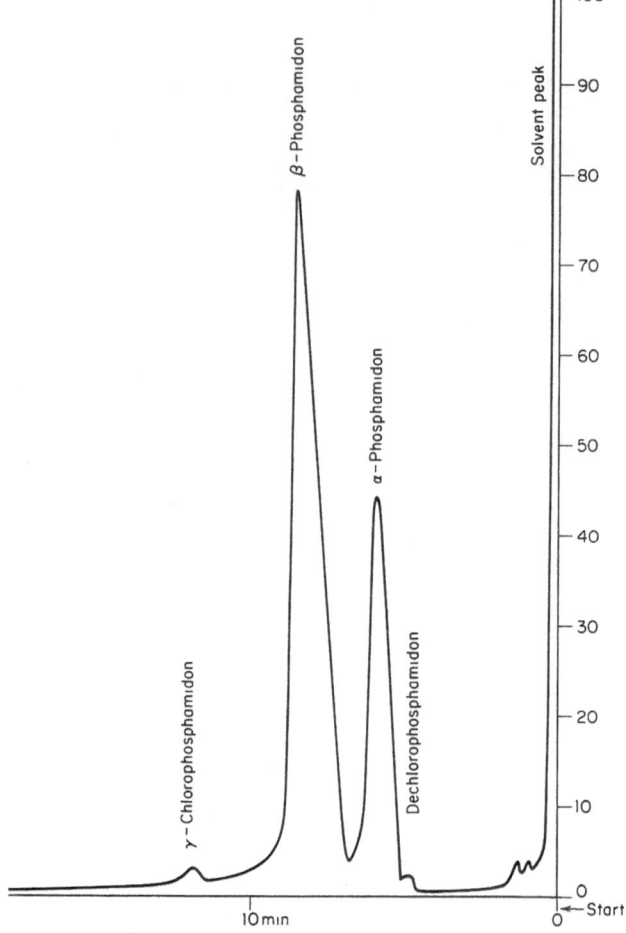

Fig. 2. Gas chromatogram of phosphamidon: 6 μl. of a 12 percent solution in benzene. Glass column 4′, diameter 3/8″, 15 percent SE on Embacel 60/180; column temperature 185° C., injector temperature 220° C., thermal conductivity cell, and helium carrier gas 75 ml./min.

ture should not exceed 185° C. The injector must be made of glass, or decomposition will occur in the temperature range between 200° and 210° C.

Inject a few ml. of phosphamidon in anhydrous isopropanol or benzene. Tributylphosphate or *n*-docosane may be used as internal standard. Tributylphosphate is eluted before α- and β-phosphamidon, followed by *n*-docosane.

The separation of the two isomers is good, α-phosphamidon appearing before the β-isomer. γ-Chlorophosphamidon and dechlorophosphamidon can be also separated by this method. Because response factors can differ considerably, frequent calibration with substances of known composition is necessary.

h) Chromatographic separation

Paper chromatographic separation (ANLIKER et al. 1961) has now largely been replaced by thin-layer chromatographic methods. The separation of the active substance from by-products or the isolation of phosphamidon from technical formulations may be achieved on thin-layer plates coated with silica gel SL-254 Merck (0.25-cm. layer). The solvent system used is a mixture of methylene chloride, ethyl acetate, and tert.-butanol in the proportions of 6 : 3 : 1 v/v. The chamber saturation system of development is used. Phosphamidon may be initially extracted from technical formulations with acetone.

The identification is made in UV-light. For the quantitative determination the silica gel is removed from the plate and hydrolyzed in a test tube. The determination, based on the diethylamide moiety (see colorimetric method), is done colorimetrically. An alternative method involves the quantitative determination of phosphorus with ammonium molybdate or vanadomolybdate after hydrolysis of the phosphamidon molecule.

The two isomers of phosphamidon are practically inseparable by TLC methods.

i) Enzymatic method

An automated enzymatic method for determining anticholinesterase insecticides, including technical grade phosphamidon, has been described by Voss (1969). This method has also been used to determine the stability of technical formulations, and various other data including solubility, distribution coefficient, and release were obtained.

IV. Discussion

Being comparatively simple, the iodometric method is widely used for the determination of active substances and formulations. The argentometric method is successful when disturbing by-products are present.

TLC separation followed by hydrolysis and colorimetric determination of phosphorus or diethylamine is a delicate method and must be done by an experienced analyst.

The gas chromatographic method is suitable for the determination of technical grade phosphamidon as well as for the analysis of formulations, from which it may be isolated by extraction. To obtain good reproducibility over a long range the response factors have to be checked frequently.

The accuracy of the automatic enzymatic determination is about as good as that of the chemical method. Twenty samples may be determined in one hour.

Acknowledgement

I wish to thank my colleagues H. Abegg, A. Becker, W. Heizler, and J. Rosales for their valuable contributions to this work.

Summary

Various methods for identification and quantitative determination of phosphamidon as active substance or in formulations have been described in the present chapter. The iodometric method is widely used in the analytical laboratory of CIBA, because it is relatively simple and suitable for routine analyses. If interfering by-products are present in the technical grade material, the argentometric method is recommended. Further procedures are based on a reaction between mercaptan and phosphamidon, and on a colorimetric determination of diethylamine. Physicochemical methods, such as infrared spectroscopy, nuclear magnetic resonance, and gas chromatography have also been applied, particularly in connection with investigations on *cis-trans* isomerism. An automated enzymatic method based on cholinesterase inhibition was found to be useful for certain types of routine tests.

Résumé *
Méthodes analytiques pour le phosphamidon

Dans ce chapitre différentes méthodes d'identification et de détermination quantitative du phosphamidon ont été décrites. Dans le Laboratoire Analytique de CIBA Société Anonyme une méthode iodométrique est souvent employée, car elle est simple et convient aux analyses de routine. La méthode argentimétrique est utilisée pour les préparations techniques qui contiennent des produits gênants. D'autres procédés sont basés, d'une part sur une réaction entre mercaptane et phosphamidon, d'autre part sur une détermination colorimétrique du diéthylamine. Les méthodes physicochimiques, telles que la spectrophotométrie infrarouge, la résonnance magnétique nucléaire et la chromatographie en phase gazeuse ont aussi été utilisées, particulièrement lors de l'étude des isomères *cis-trans*. Une méthode enzymatique automatique s'est révélée utile pour certains types d'analyses de routine.

Zusammenfassung **
Analytische Methoden zur Bestimmung von Phosphamidon

Im vorliegenden Kapitel wurden verschiedene Methoden zur Identifizierung und quantitativen Bestimmung von Phosphamidon beschrieben. Im

* Traduit par L. P. LANG.
** Übersetzt vom Autor.

Analytischen Laboratorium der CIBA wird eine jodometrische Methode häufig angewendet, da sie einfach ist und sich für Routineanalysen eignet. Die argentometrische Methode dagegen wird für technisches Material benutzt, das störende Nebenprodukte enthält. Weitere Verfahren beruhen auf einer Reaktion zwischen Mercaptan und Phosphamidon, sowie auf einer colorimetrischen Bestimmung von Diäthylamid. Darüberhinaus wurden, besonders im Zusammenhang mit Untersuchungen zur *cis-trans*-Isomerie, physikalisch-chemische Methoden benutzt, wie z. B. Infrarotspektroskopie, kernmagnetische Resonanz und Gaschromatographie. Für gewisse Routine-Untersuchungen eignet sich eine automatische Methode, die auf der Cholinesterasehemmung basiert.

References

ANLIKER, R., E. BERIGER, M. GEIGER, and K. SCHMID: Über die Synthese von Phosphamidon und seinen Abbau in Pflanzen. Helv. Chim. Acta 44, 1622 (1961).

California Chemical Co. (now. *Chevron Chem. Co.*), Ortho Division, private communication.

CROSBY, N. T., and E. Q. LAWS: The use of infrared spectroscopy in the analysis of pesticide residues. Analyst 89, 319 (1964).

GETZ, M. E., and R. R. WATTS: Application of 4-(p-nitrobenzyl)-pyridine as a rapid quantitative reagent for organophosphate pesticides. J. Assoc. Official Anal. Chemists 47, 1094 (1964).

LICHTENTHALER, F. W.: The chemistry and properties of enol phosphates. Chem. Rev. 61, 607 (1961).

MADER, W. J., and R. R. BUCK: Colorimetric determination of cortisone and related ketol steroids. Anal. Chem. 24, 666 (1952).

PACK, D. E., J. N. OSPENSON, and G. K. KOHN: Phosphamidon. In G. ZWEIG, ed.: Analytical methods for pesticides, plant growth regulators, and food additives, vol. II, p. 375. New York-London: Academic Press (1964).

RUZICKA, J., J. THOMSON, and B. B. WHEALS: The gas chromatographic examination of organophosphorous pesticides and their oxidation products. J. Chromatog. 30, 92 (1967).

VOEGELI, P., and F. CHRISTEN: Bestimmung von organisch gebundenem Halogen mit Piperidin. Z. Anal. Chem. 233, 175 (1968).

VOSS, G.: Cholinesterase inhibition autoanalysis of insecticidal organophosphates and carbamates. J. Assoc. Official Anal. Chemists 52, 1027 (1969).

Chapter 3

Formulations of phosphamidon

By

M. Geiger

Contents

I. Introduction

Dimecron is the trade mark of *CIBA Ltd.* for formulations containing the active ingredient phosphamidon. The concentration of the formulations is always defined as the total active ingredient content, *i. e.*, α- and β-phosphamidon plus γ-chlorophosphamidon and dechlorophosphamidon (see Chapter I). In all liquid formulations, the active ingredient content is given in percent weight per volume; in the solid formulations it is given in percent weight per weight. As phosphamidon is water soluble and its liquid formulations are miscible with water, they are simply called Dimecron 20, 50, etc., without indication of the type of formulation.

II. General considerations

Being a liquid which is soluble in water and most organic solvents, phosphamidon is readily formulated as water-miscible concentrates. On the other hand, it gives difficulty in wettable powders, dusts, and granular formulations.

a) Solubility characteristics of phosphamidon

Phosphamidon is a polar compound which mixes with water in all proportions. The addition of salts (e. g., sodium chloride) reduces its water solubility quite considerably (see Table I), and this is important when it is

Table I. *Solubility of phosphamidon and dicrotophos in sodium chloride solutions of different concentrations at 20° C.*

NaCl (g./100 ml. solution)	Phosphamidon solubility (g./100 ml.)	Dicrotophos solubility (g./100 ml.)
3	miscible	miscible
4	11	miscible
6	6	miscible
8	4	miscible
10	3	miscible
15	1.8	miscible
18	—	70
20	—	56
22	—	47
24	—	40
26	—	33

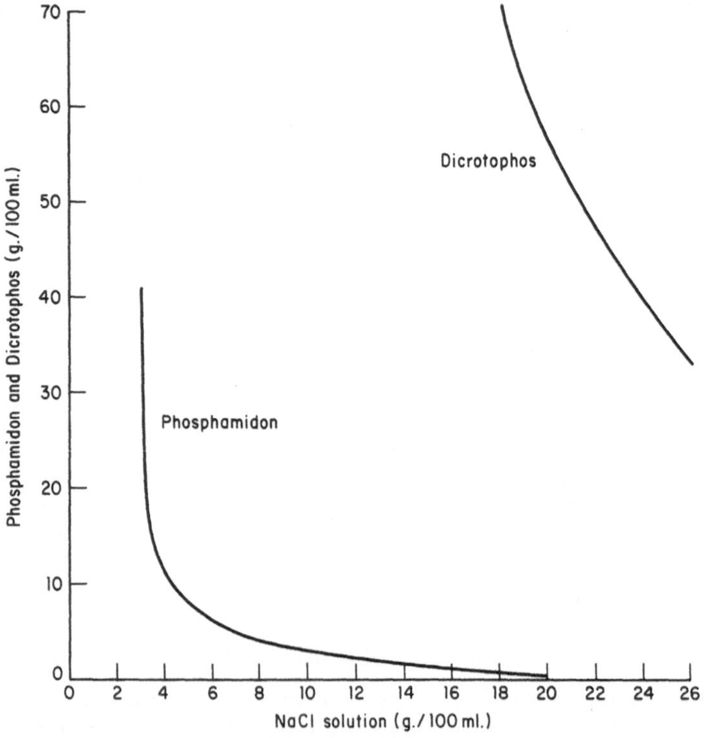

Fig. 1. Solubility of phosphamidon and dicrotophos in sodium chloride solutions of different concentrations

mixed with protein hydrolysate insecticide baits. Dicrotophos is more suitable for such baits, as it is more polar, and its aqueous solutions are more tolerant of higher salt concentrations (CIBA 1969 a; Figure 1).

Phosphamidon is soluble in all proportions in alcohols, glycols, ketones, ethers, esters, and aromatic hydrocarbons, but less soluble in saturated hydrocarbons (e. g., hexane and cyclohexane). High-boiling saturated kerosenes dissolve only small quantities of phosphamidon, and mineral oils even less (see Table II).

Table II. *Solubility of technical grade phosphamidon in different solvents at 20° C.*

Solvent	Solubility at 20° C. (% w/v)
Water	miscible
Alcohols	miscible
Glycols	miscible
Ketones	miscible
Ethers	miscible
Esters	miscible
Chlorinated hydrocarbons	miscible
Aromatic hydrocarbons	
Xylene	miscible
Shellsol R [a]	miscible
Solvent 200 [b]	miscible
Saturated hydrocarbons	
Hexane	2
Shellsol T [c]	2
Cyclohexane	2.7
Dutrex 3 SP [d]	11
Carnea Oil 21 [c]	1
Cottonseed oil	1

[a] Trade mark of *Shell* for petroleum solvent with boiling range 210° to 270° C., 83 percent aromatics.
[b] Trade mark of *Esso-Standard* for petroleum solvent with boiling range 230° to 270° C., 98 to 100 percent aromatics.
[c] Trade mark of *Shell* for kerosene with boiling range 180° to 210° C., 0.5 percent aromatics.
[d] Trade mark of *Shell* for processed petroleum extract of high boiling range and high aromatic content.
[e] Trade mark of *Shell* for mineral oil with sulfonation index of about 70.

b) Chemical reactivity of phosphamidon with formulation auxiliaries

Like other vinyl phosphates, phosphamidon is subject to degradation, particularly by hydrolysis (ANLIKER *et al.* 1961). Other mechanisms come into play when vinyl phosphates are stored at high temperatures. These were described by BROWN *et al.* (1966) for monocrotophos and dicrotophos.

Table III. *Decomposition of technical grade phosphamidon at different temperatures*

Storage temperature (°C.)	Phosphamidon contents after x weeks (%)					
	0×	2×	5×	8×	10×	24×
20	90	—	90.5	—	89	90
25	—	—	89.5	—	89	89
50	—	—	89	87	—	81
70	—	88	85.5	84	—	78

Table IV. *Decomposition of 20 and 50 percent technical grade phosphamidon solutions in different hydroxylic solvents*

Solvent	Phosphamidon contents after x months (%)											
	20% solutions						50% solutions					
	35° C.			50° C.			35° C.			50° C.		
	1×	2×	3×	1×	2×	3×	1×	2×	3×	1×	2×	3×
Methanol	17	13	10.5	10	4	—	46	37.5	37	27.5	11	—
Ethanol	18.5	15.5	14.5	13	6	—	47.5	47	39	37.5	23	—
Ethanol+ 5% water	17	13	10.5	11	4	—	45.5	37.5	33	29.5	15	—
Isopropanol	19	18	17.5	17.5	14	10	49	47.5	46.5	45.5	38	35.5
Cellosolve	20	19.5	19.5	18	16.5	14.5	49	49	47	45.5	41	37
Cellosolve +1% water	19.5	18.5	18	17.5	15	12.5	49.5	47.5	46	45	38	31.5

The breakdown rates of technical grade phosphamidon at temperatures in the range 22° to 70° C. are given in Table III (*CIBA* 1969 b). As might be expected the decomposition rate increases with temperature.

The effect of solvents on the stability of phosphamidon depends on their chemical nature. Alcohols react with phosphamidon, and the reaction rate is greatest with primary alcohols; the rate decreases in the order methanol, ethanol, *n*-propanol. The primary alcohol ethoxyethanol (Cellosolve) behaves very favourably. Secondary alcohols (e. g., isopropanol and *sec.*-butanol) are much less reactive. Table IV shows the stability of phosphamidon in different hydroxylic solvents (*CIBA* 1969). It is obvious that water in the solvent accelerates the decomposition of phosphamidon, but no quantitative relationship was established.

Organic solvents other than alcohols are less reactive, and are preferable from this standpoint. Solubility of the solvent in water is an important requirement for the preparation of spray mixtures with Dimecron formulations (see subsection *c*, below), and in stability tests we therefore concentrated on water-miscible and at least partially water-soluble solvents. Some results are given in Table V.

Table V. *Stability of phosphamidon in some non-hydroxylic solvents*

| Solvent | Phosphamidon in 50% solutions after x months (%) | | | | | |
| | 35° C. | | | 50° C. | | |
	1×	3×	6×	1×	3×	6×
Xylene	—	—	—	50	50	49
Xylene + 10% emulsifier a	—	—	—	44	42.5	41
Pentoxone b	—	—	44	—	—	20
DAC c	—	—	50	—	—	46
Cellosolve acetate d	50	50	50	50	50	—
Triethyl phosphate	—	—	50	—	—	46

a Toximul MP = blend of anionic with nonionic emulsifiers.
b Trade mark of *Shell* for 2-methyl-2-methoxy pentanone-4.
c Diethyl carbitol = diethyleneglycol diethyl ether.
d Acetate of ethylene glycol monoethyl ether.

The influence of wetting agents and emulsifiers on the storage stability of phosphamidon depends on their chemical nature. Nonionic agents are practically without effect, but anionic compounds such as calcium sulfonates, which are used to bring about spontaneous emulsification of emulsifiable concentrates, greatly reduce the stability of phosphamidon formulations (see Table V).

The reactivity of phosphamidon with solid carrier materials such as mineral or synthetic inorganic and organic products, is a severe problem. On the one hand there are catalytically active, so-called hot spots in these materials which induce decomposition on storage, while on the other hand the high sorptivity of such carrier materials for active ingredients is normally tied with high water adsorptivity. Well-dried, precipitated silica has proved to be the best carrier material for phosphamidon, but care must be taken to exclude water during manufacture and storage.

c) Points to be borne in mind when choosing solvents for the formulation of Dimecron

Different factors are to be considered in choosing a proper solvent for Dimecron formulations. The main ones are listed below without respect to their relative importance:

1. Chemical compatibility with phosphamidon
2. Water solubility
3. Compatibility with packing materials
4. Price and availability

None of the solvents in current use fulfills all these requirements to a satisfactory degree, and we must, therefore, find a reasonable compromise. This is, of course, difficult, because everybody looks at this problems from an individual aspect.

1. Chemical compatibility. — This aspect was covered in subsection *b*. The results of the stability tests clearly indicate that alcohols should be avoided. However, the low price and ready availability of isopropanol throughout the world have favoured its use, and experience over a period of ten years has shown it to be an acceptable solvent. Chemical breakdown may be compensated by a certain surplus of the active ingredient in the formulation. Low percentage formulations with 10 to 20 percent active ingredient are normally used in cool climates and the more concentrated ones, which are distinctly more stable, in warmer climates.

A certain amount of chemical breakdown of the isopropanol formulations does not affect the preparation of the spray liquid.

2. Water solubility. — One of the most important features of phosphamidon is its miscibility with water in all proportions. Clearly, therefore, water-soluble solvents should be used in the formulation of Dimecron. This means that emulsifiers can be avoided, which brings a range of advantages: there are no emulsification and emulsion stability problems; fruit russeting, often caused by emulsions, does not occur; and the compatibility of Dimecron with other pesticides in tank-mix combinations, emulsions, or suspensions is usually good.

If mixed formulations of phosphamidon with other insecticides (e. g., fenitrothion or DDVP) which are only slightly water-soluble are required, it is necessary to use less polar solvents and emulsifiers to form emulsifiable concentrates. Certain solvents bring serious packaging problems, and in some cases glass containers are the only possibility.

3. Compatibility with packaging materials. — Steel, tinplate, and aluminium containers are not acceptable, because the acid by-products, always present in small quantities, are corrosive. Special lacquers, applied in two layers with subsequent curing, may be used. Unichrome B 124 is chemically resistant, but brittle.

The most convenient packaging material is polyethylene. Phosphamidon is sufficiently polar to be unable to penetrate polyethylene of the generally available quality. The non-permeability of polyethylene by polar solvents is an important reason for preferring them to non-polar solvents which do permeate polyethylene.

4. Price and availability. — Commercial considerations lead to the conclusion that, for low concentrate Dimecron formulations, isopropanol is the most economic solvent. In countries where flash point specifications prevent the use of this solvent (flash point 11° C. in Pensky Martens closed cup tester) the more expensive Cellosolve (flash point 40° C.) may be used. Where the stability requirement is more stringent, Cellosolve acetate is an alternative. The most convenient solvent is chosen in the light of the specific requirements. For the standard commercial formulations, however, 99 percent pure isopropanol, with a water content of less than 0.2 percent, is used. This quality is available from many manufacturers throughout the world.

III. Commercial formulations

a) Liquid formulations

The liquid formulations currently available are Dimecron 20, 50, and 100. These are solutions of technical grade phosphamidon in isopropanol. Their composition is shown in Table VI and their chemical stability at different temperatures in Table VII (*CIBA* 1969 b). The formulations are based on the 92 percent active ingredient content of technical grade phosphamidon, which for warning purposes is coloured by addition of 0.06 percent of Violet 5 BO oleate. The formulation procedure is extremely simple, the two liquid components being simply mixed in a stirring kettle.

Table VI. *Composition of liquid Dimecron formulations*

Formulation	Phosphamidon		Isopropyl alcohol (kg./100 l.)	Specific gravity (kg./l.)
	Pure (kg./100 l.)	Technical 92% (kg./100 l.)		
Dimecron 20	20.0	21.8	64.6	0.864
Dimecron 50	50.0	54.5	43.9	0.984
Dimecron 100	100.0	109.0	8.6	1.176

Table VII. *Storage life of commercial Dimecron formulations*

Preparation	Storage temp. (°C.)	Phosphamidon content as percent of initial concentration after x weeks				
		2×	5×	8×	10×	24×
Dimecron 100	20	—	99.5	—	98.5	100
	35	—	99.5	—	98.5	99
	50	—	98.5	97	—	94
	70	96	88.5	82	—	68.5
Dimecron 50	20	—	100	—	98	98
	35	—	98	—	94	87
	50	—	90.5	82	—	73
	70	79	50	27	—	12
Dimecron 20	20	—	100	—	95	97
	35	—	95	—	90	80
	50	—	83	73	—	50
	70	68	40	20	—	13

Quality control is achieved by determining the specific gravity and the active ingredient content (iodometric assay method, *CIBA* 1969 c) (see chapter 2).

Phosphamidon (Dimecron 100) can be applied by ultra-low-volume (ULV) techniques. If ULV formulations of lower concentration (e. g., 25 or 50 percent w/v) are required, technical grade phosphamidon may be diluted with high boiling aromatic hydrocarbons or other suitable solvents, but this involves a certain risk of phytotoxicity, and it is advisable to test such formulations on the crop in question before extended use.

b) Dry Dimecron formulations

Dimecron 50 WP is prepared by impregnating 44 parts of specially dried silica powder with 56 parts of 92 percent technical grade phosphamidon (containing 0.25 percent Violet 5 BO oleate). It is a free-flowing powder, which disperses readily in water, the phosphamidon dissolving while the silica slowly settles out. As far as the active ingredient is concerned, the formulation is in fact a soluble powder. The manufacturing equipment is a high energy mixer with a high speed rotary spraying or pressure spraying device. The active ingredient content of the product is checked by the iodometric procedure (*CIBA* 1969 c) (see chapter 2) using a solution of 3 g. of WP 50 in 200 ml. of water.

Dimecron dusts are prepared by mixing the 50 WP with suitable dust carriers. Such formulations normally have an active ingredient content of one to five percent and may be combined if necessary with other pesticide dusts. They tend to be unstable in storage, and in each case a suitable mineral carrier of local origin must be found for the local formulation.

Dimecron granules are not yet commercially available, but fast and slow release types are being investigated.

IV. Storage life of commercial Dimecron formulations

As a rule it can be assumed that concentrated preparations have a longer storage life than dilute ones, and that the storage life is shortened as the temperature rises. Table VII (*CIBA* 1969 b) shows the phosphamidon content of a series of Dimecron formulations after storage at various temperatures as a percentage of the original phosphamidon content. It is clear that the stability depends on the concentration of the formulation and also on the temperature of storage.

There is no separation at temperatures down to 20° C. below zero.

V. Preparation of sprays, compatibility, and wetting properties

When poured into water, liquid Dimecron formulations immediately dissolve to give clear spray solutions of pale violet colour. In this manner spray solutions of any desired concentrations can be prepared, and this is of particular importance in ULV work.

A suspension of Dimecron 50 WP is prepared by stirring the powder into water. Lumps disappear immediately when stirred. The phosphamidon is completely dissolved out of the powder by the water, while the dye remains for the most part in the carrier material. The dark-coloured carrier settles at the bottom of the spray solution, but is so fine and loose that it does not interfere with the spraying.

Since neither the liquid Dimecron formulations nor the WP 50 contain emulsifying agents, and phosphamidon is nonionic, these formulations may be mixed with virtually all plant pesticides in the form of aqueous solutions, emulsions, or suspensions. Strongly alkaline preparations are an exception, destroying phosphamidon, like most organophosphorus compounds, by hydrolysis. Alkaline products of this type, however, have virtually fallen into disuse. A further reservation regarding the compatibility of Dimecron with other preparations concerns the combination of liquid Dimecron with highly concentrated emulsions and suspensions, in which case the equilibrium of the dispersion may be irreversibly upset by the high concentration of phosphamidon or of isopropanol. There is no generally applicable rule for calculating compatibility in such cases, but there is usually an answer for each individual problem.

Practice has shown that Dimecron spray solutions have sufficient wetting power for most purposes, and in general it is not advisable to add a wetting agent because it causes excessive run-off (e. g., in deciduous fruit spraying, where tank-mixing of Dimecron with fungicides is very common). Where the crop to be treated is particularly difficult to wet, a wetting agent may be used. Wetting agents based on alkylphenol polyglycol ethers are among the many which are suitable.

Summary

Since phosphamidon is soluble in water and water-miscible organic solvents in all proportions, it is readily formulated as a water-miscible concentrate. Isopropanol, being inexpensive and widely available, was found to be acceptable for formulations, although alcohols, particularly primary ones, show a certain reactivity with phosphamidon. The use of isopropanol, furthermore, permits phosphamidon to be packed in polyethylene containers, which material is not penetrated by polar solvents. With regard to solid preparations, well-dried silica has proved to be useful as a carrier, if water is excluded during manufacture and storage.

Dimecron 20, 50 and 100 solutions of technical grade phosphamidon in isopropanol are the major commercial liquid formulations. The latter one appears to be the most stable formulation, and it is also being used for ultra low-volume applications. Dimecron 50 wettable powder represents the major dry formulation. It is miscible with suitable carriers to yield dusts of one-to-five percent active ingredient, which can be combined with other pesticide dusts. Both liquid and dry formulations are compatible with most plant pesticides in the form of aqueous solutions, emulsions, or suspensions.

Résumé *

Différentes préparations à base de phosphamidon

Le phosphamidon étant soluble en toute proportion, dans l'eau et dans les solvants polaires, on peut facilement le préparer en concentré miscible à l'eau. L'isopropanol étant bon marché et largement répandu, il se prête bien aux préparations, quoique les alcools primaires présentent une certaine réactivité vis-à-vis de l'insecticide. L'emploi d'isopropanol permet en outre le stockage en récipients de polyéthylène, matériau qui n'est par les solvants polaires. Pour les préparations solides, la silice sèche s'est révélée être un support utile, lorsqu'on évite l'humidité pendant la fabrication et le stockage.

Les principales préparations liquides commerciales sont le dimecron 20, 50 et 100. Celles-ci sont des solutions de phosphamidon technique dans l'isopropanol. Le dimecron 100 est la préparation la plus stable, qui se prête également aux applications en volumes "ultra low". Le dimecron 50 WP constitue la principale forme solide. On peut le mélanger à des supports convenables, ce qui permet l'obtention de poussières contenant 1 à 5 pour cent de produit actif. Ces dernières peuvent à leur tour être combinées avec d'autres poussières de pesticides. Les préparations liquides et solides sont compatibles avec la plupart des pesticides végétaux, sous forme de solutions aqueuses, d'émulsions ou de suspensions.

Zusammenfassung **

Formulierungen des Phosphamidons

Da Phosphamidon sich in jedem Verhältnis mit Wasser oder wasserlöslichen organischen Lösungsmitteln mischen läßt, kann man es gut als wassermischbares Konzentrat formulieren. Isopropanol, ein billiges und vielerorts zu erhaltendes Lösungsmittel, ist für Formulierungen besonders geeignet, obwohl Alkohole eine gewisse Reaktionsfähigkeit mit Phosphamidon zeigen. Die Verwendung von Isopropanol gestattet ferner eine Verpackung in Polyäthylen-Behältern; dieser Kunststoff wird durch polare Lösungmittel nicht angegriffen. Für feste Formulierungen hat sich trockene Kieselsäure bewährt, sofern während der Produktion und Lagerung Feuchtigkeit ausgeschlossen wird.

Die wichtigsten kommerziellen Flüssigkeitsformulierungen sind Dimecron 20, 50 und 100, Lösungen von technischem Phosphamidon in Isopropanol. Dimecron 100 ist die stabilste Formulierung; sie wird auch für „ultra-low-volume"-Applikationen verwendet. Die wichtigste feste Formulierung wird durch Dimecron 50 WP repräsentiert. Es ist mit bestimmten Trägermaterialien mischbar, wodurch sich Staubformulierungen mit ein bis fünf Prozent Wirkstoff herstellen lassen, die ihrerseits wieder mit anderen

---

* Traduit par J. P. Lang.
** Übersetzt vom Autor.

Pestizid-Stäuben kombiniert werden können. Mit wäßrigen Lösungen, Emulsionen oder Suspensionen der meisten Pflanzenschutzmittel zeigen flüssige und feste Formulierungen des Phosphamidons eine gute Verträglichkeit.

References

ANLIKER, R., E. BERIGER, M. GEIGER, and K. SCHMID: Über die Synthese von Phosphamidon und seinen Abbau in Pflanzen. Helv. Chim. Acta 44, 1622 (1961).

BROWN, N. P. H., A. S. FORSTER, and C. G. L. FURMIDGE: Stability of agricultural chemicals. Hydrolytic and thermal stabilities of phosphorylated crotonamides. J. Sci. Food. Agr. 17, 510 (1966).

CIBA, Ltd., Agrochemical Division, Basle, Switzerland: TA 749: Stabilität von Dimecron-Lagerproben. Unpublished report (1965).

— Löslichkeit von Phosphamidon und Dicrotophos in Kochsalzlösungen verschiedener Konzentration. Unpublished report (1969 a).

— Lagerfähigkeit von Phosphamidon und Dimecron 20, 50, 100 in Isopropanol. Unpublished report (1969 b).

— Methods of analysis: Dimecron 100, Code-No. E-95 5/2, 10. 9. 1968; Dimecron 50, Code-No. E-102 R/1, 22. 1. 1968; Dimecron 20, Code-No. E-171 R/1, 20. 5. 1969. Methods available on request (1969 c).

Chapter 4

The metabolism of phosphamidon in plants and animals

By

H. Geissbühler, G. Voss, and R. Anliker

Contents

I. Introduction

By the time phosphamidon was developed as an insecticide, it was generally recognized that organophosphorus compounds are subjected to chemical and/or biochemical (enzymatic) transformation processes in plant, mammalian, and insect tissues (O'Brien 1960, Heath 1961). Biologically speaking, these transformation processes represent either activation or degradation. With regard to organophosphate insecticides, activation means conversion of the parent compound to a metabolite with increased anticholinesterase activity, whereas degradation means conversion to metabolites which no longer inhibit cholinesterases, or which do so to a much smaller degree (O'Brien 1967, Fukuto and Metcalf 1969). The toxicological significance of organophosphorus metabolites normally parallels their anticholinesterase activity, although this may be masked or offset by the ease with which a particular anticholinesterase agent is degraded within the animal body.

Phosphamidon, being a phosphate-type compound (in contrast to phosphorothioate insecticides), was immediately recognized to be itself a strong inhibitor of cholinesterases (Jacques and Bein 1960). The principal aim of the metabolite studies described below was, therefore, to examine plant and animal tissues for the presence of additional anticholinesterases derived from the parent compound by chemical and/or enzymatic conversion. Furthermore, hydrolytic breakdown of phosphamidon was to be expected to yield chlorinated acetoacetic acid amides, the persistence and toxicological significance of which had to be evaluated. For this reason, the *CIBA* scientists who were first involved in metabolism experiments decided to forego the commonly used ^{32}P-labeling of organophosphorus insecticides (Casida 1962) and to work with a phosphamidon preparation labeled with ^{14}C in the vinyl-portion of the molecule (Anliker *et al.* 1961 a). The label was introduced both at the methylvinyl and at the carbonyl positions of the compound to yield radioactive phosphamidon with the high specific activity of 3.6 mc/mM (for details of the radiosynthesis see Chapter 1).

$$\text{CH}_3\text{O} \diagdown \underset{\underset{\text{CH}_3\text{O}}{}{\overset{\displaystyle \text{O}}{\diagup}}}{\text{P}} \diagdown \text{O}\text{—}^*\underset{\underset{\text{CH}_3}{|}}{\text{C}}=\underset{\underset{\text{Cl}}{|}}{\text{C}}\text{—}^*\overset{\overset{\displaystyle \text{O}}{\diagup}}{\text{C}} \diagdown \underset{\underset{\text{C}_2\text{H}_5}{}}{\overset{\displaystyle \text{C}_2\text{H}_5}{\text{N}}} \qquad\qquad [\text{I}]$$

From the general reactions involved in organophosphorus metabolism, which were recognized at the time the phosphamidon experiments were started, it was to be expected that the insecticide would be hydrolyzed at the P-O-vinyl and P-O-methyl linkages (O'Brien 1960). The fate of the alkylated amide moiety was not so easily predictable. However, oxidative *N*-dealkylation of substituted amines and amides had already been demonstrated for various compounds other than pesticides (Williams 1959), and evidence had been presented for the oxidation of substituted phosphoroamidates [for example, octamethylpyrophosphoramide (OMPA, Schradan)] to form the corresponding *N*-oxide and/or *N*-methylol derivatives, both of which were observed to be strong anticholinesterases (Tsuyuki *et al.* 1955, Spencer *et al.* 1957). Further metabolic reactions of phosphamidon which had to be considered were deamidation and/or dechlorination of the parent compound or its degradation products (O'Brien 1960).

The present chapter is particularly concerned with the pathways of biochemical transformation of phosphamidon in plant and animal tissues, and the techniques applied for separating, isolating, and identifying the observed metabolites. Some purely chemical aspects of the synthesis of radio-labeled compounds and metabolites, and methods for their separation, were dealt with in Chapter 1. The dynamics of the degradation of phosphamidon and its toxic metabolites in plants are described in Chapter 8. The way in which methods of residue analysis are designed to take into account the effects of plant metabolism is discussed in Chapter 7. The toxicological implications and public health aspects of phosphamidon metabolism in plants and animals are dealt with extensively in Chapter 5.

II. Metabolism in plants

The first investigations of the metabolism of phosphamidon in plants were performed by *CIBA* scientists and reported in 1961 (Anliker *et al.*

1961 b). When examining various plant materials for residues of the insecticide, these authors observed that heavily sprayed leaves of French beans *(Phaseolus vulgaris)* contained one major and two minor transformation products which separated from the parent compound on paper chromatograms (Fig. 1). The color spot test with tetrazolium blue (see Chapter 1) was positive for all three metabolites, indicating that major portions of the original phosphamidon structure were still intact in their molecules.

The main metabolite was isolated from about three kg. of bean material, concentrated, and purified. Identification of the colorless oil by infrared spectroscopy and elementary analysis as well as comparison with a synthetic standard showed it to be a *cis-trans-* mixture of 2-chloro-2 ethyl-carbamoyl-1-methylvinyl dimethylphosphate (*N*-desethyl phosphamidon):

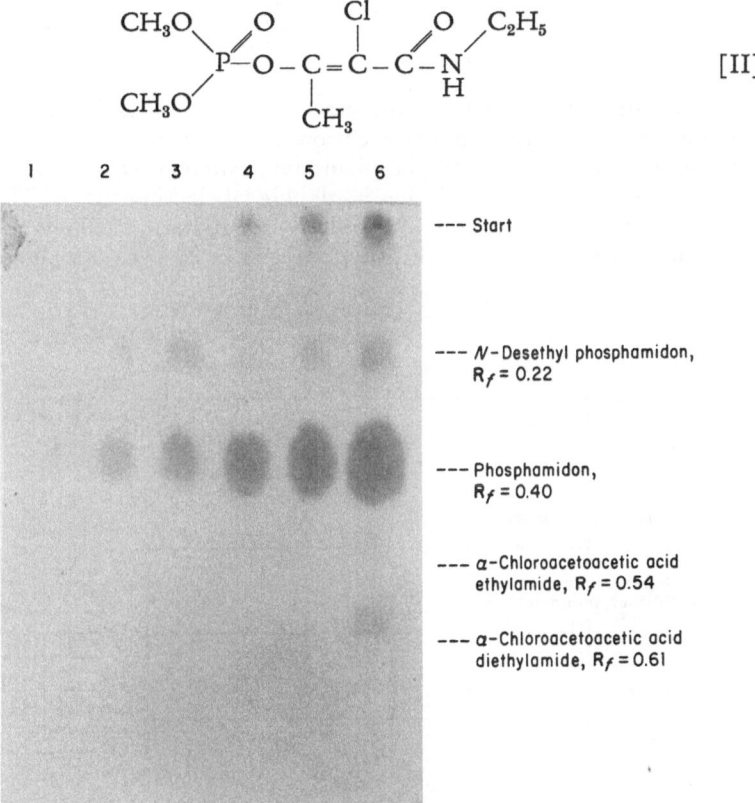

Fig. 1. Transformation products of phosphamidon in insecticide-treated French bean leaves as detected by paper chromatography. Plant extract prepared in accordance with the phosphamidon residue method described by *CIBA* (1964 e). Solvent system B₁ as described by BUSH (1952). Chromogenic agent: blue tetrazolium. Spots no. 1 to 3 reference compounds, 4 to 6 actual plant extracts

The two minor metabolites extracted from bean leaves were suspected to be phosphorus-free hydrolysis products of phosphamidon and/or *N*-desethyl phosphamidon. Their partitioning behaviour in several solvent systems and their paper chromatographic properties corresponded with those of synthetic samples of α-chloroacetoacetic acid diethylamide and α-chloroacetoacetic acid ethylamide, respectively (Fig. 1):

$$CH_3 - \overset{\displaystyle O}{\underset{}{C}} - \overset{\displaystyle Cl}{\underset{}{C}} - \overset{\displaystyle O}{\underset{}{C}} - N\overset{\displaystyle C_2H_5}{\underset{\displaystyle C_2H_5}{}} \qquad [III]$$

$$CH_3 - \overset{\displaystyle O}{\underset{}{C}} - \overset{\displaystyle Cl}{\underset{}{C}} - \overset{\displaystyle O}{\underset{}{C}} - N\overset{\displaystyle C_2H_5}{\underset{\displaystyle H}{}} \qquad [IV]$$

To confirm the identity of the described metabolites, to study the kinetics of their formation and degradation, and to examine further potential transformation pathways of phosphamidon, Anliker *et al.* (1961 b) continued their experiments with the ^{14}C-double labeled insecticide. The radioactive compound was again applied to French bean seedlings either by deposition at the leaf base with a micropipette or by even distribution over

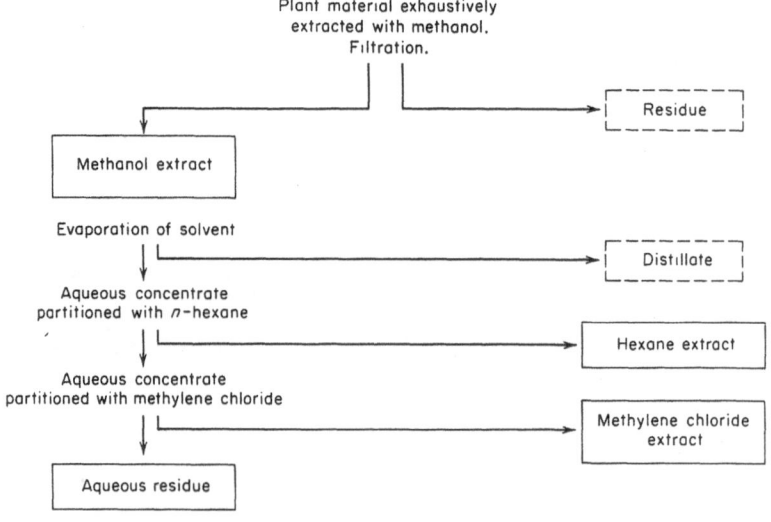

Fig. 2. Scheme of extraction, fractionation, and purification of French bean seedlings after leaf application of ^{14}C-double labeled phosphamidon. Fractions in solid frames subjected to paper chromatography with solvent systems described in Figures 1 and 3. Fractions in dotted frames free from radioactivity and therefore discarded. Method described by Anliker *et al.* (1961 b)

the entire leaf surface in the presence of a wetting agent. The concentrations used correspondend to those observed in practice and were of the order of 10 to 20 p.p.m. At different time intervals after application (one to 16 days), the plant material was homogenized and exhaustively extracted, and the extract fractionated according to the scheme presented in Figure 2.

The total radioactivity present in the four main fractions of the scheme (methanol, hexane, methylene chloride, and aqueous residue) was determined by combustion and scintillation counting. The balance of radioactivity observed in a typical experiment is summarized in Table I. These data demonstrate that the loss of total radioactivity during the 16-day observation period was relatively slight, amounting to no more than 35 percent of the original amount. This table also shows that the percentage of label present in the methylene chloride fraction, which presumably comprised the bulk of phosphamidon and N-desethyl phosphamidon, decreased quite rapidly and represented only four percent at the end of the experiment.

Table I. *Balance of radioactivity of main fractions [a] derived from French bean leaves after treatment with* ^{14}C-*double-labeled phosphamidon* [b]

Days after application	Radioactivity in percent of dose applied			
	Methanol extract	Hexane phase	Methylene chloride phase	Aqueous phase
1	91.5	0.7	68.5	12.0
2	95.6	0.8	45.4	32.2
4	82.9	0.9	17.0	62.7
8	70.5	0.6	7.9	70.2
16	73.5	0.9	4.3	66.3

[a] See Fig. 2.
[b] Adapted from ANLIKER et al. (1961 b) with permission of Helv. Chim. Acta.

Further separation of the radioactivity present in the main fractions was achieved by paper chromatography (Fig. 3). Neither the hexane nor the methylene chloride fraction revealed any major new metabolites beyond those already described. A minor peak which sporadically appeared in the organic fractions was later shown to be one of the two geometrical isomers of N-desethyl phosphamidon (ANLIKER 1963, *CIBA* 1964 a) the discovery and properties of which are described in detail in Chapter 1.

Some time after application of the labeled insecticide, the bulk of radioactivity was observed to represent polar materials which were not extracted with hexane or methylene chloride from water and which stayed at the origin with the standard chromatographic solvent system used (Table I and Fig. 3). Assuming that further transformation of phosphamidon or its metabolites might proceed by a second N-dealkylation step or by deamidation, O-desmethylation, or dechlorination, ANLIKER et al. (1961 b)

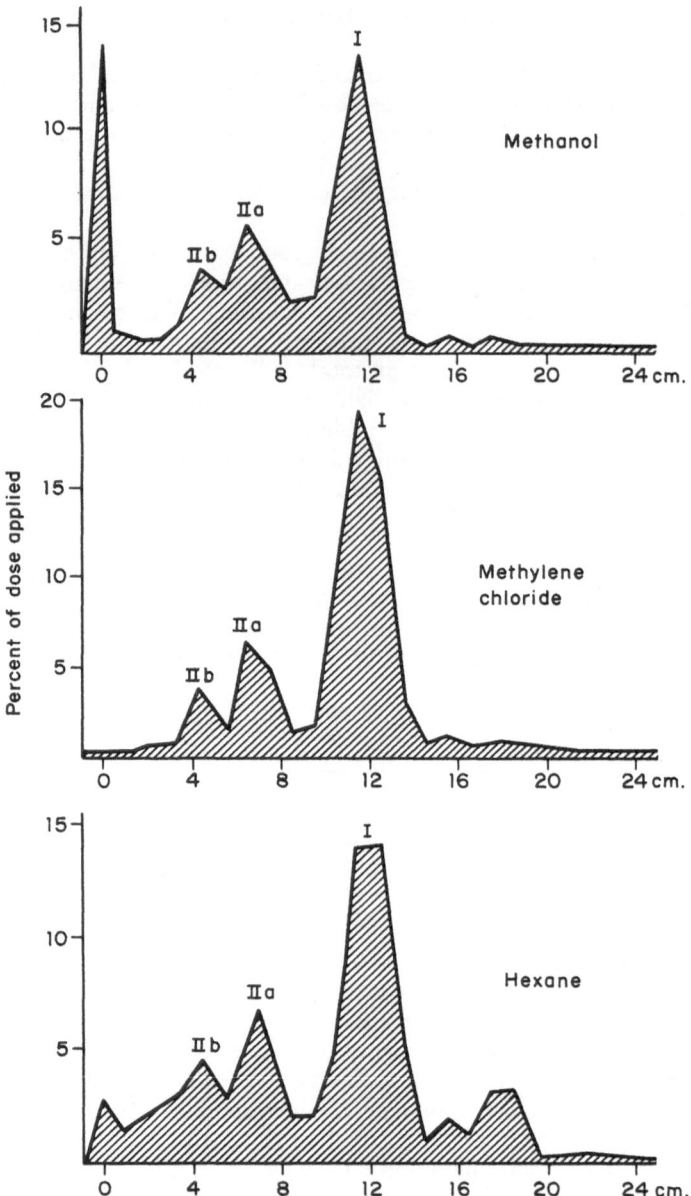

Fig. 3. Separation by reversed-phase chromatography of radioactivity recovered in main fractions (see Fig. 2) derived from French bean leaves treated with ^{14}C-double-labeled phosphamidon. Plants harvested two days after insecticide application. Solvent system B_1 according to Bush (1952): Upper phase of system petroleum ether (b. p. 90° to 95° C., 250 ml.) + toluene (250 ml.) + methanol (350 ml.) + water (150 ml.). II = phosphamidon (mixture of *cis*- and *trans*-isomers), II *a* = *cis*-isomer of *N*-desethyl phosphamidon, II *b* = *trans*-isomer of *N*-desethyl phosphamidon, III = α-chloroacetoacetic acid diethylamide, IV = α-chloroacetoacetic acid ethylamide. Adapted from ANLIKER *et al.* (1961 b)

prepared the reference compounds listed in the top part of Table II. However, neither the partitioning behaviour nor the chromatographic and electrophoretic properties of these references corresponded to those of the polar radioactive fractions which had been isolated.

Table II. *Reference compounds synthesized for comparison with radioactive fractions extracted from French bean leaves after application of ^{14}C-phosphamidon. Top part of table: potential metabolites of phosphamidon (cis- and trans-isomers) to be expected from recognized biochemical transformation reactions [a]. Bottom part of table: potential metabolites observed in model hydrolysis experiments [b]*

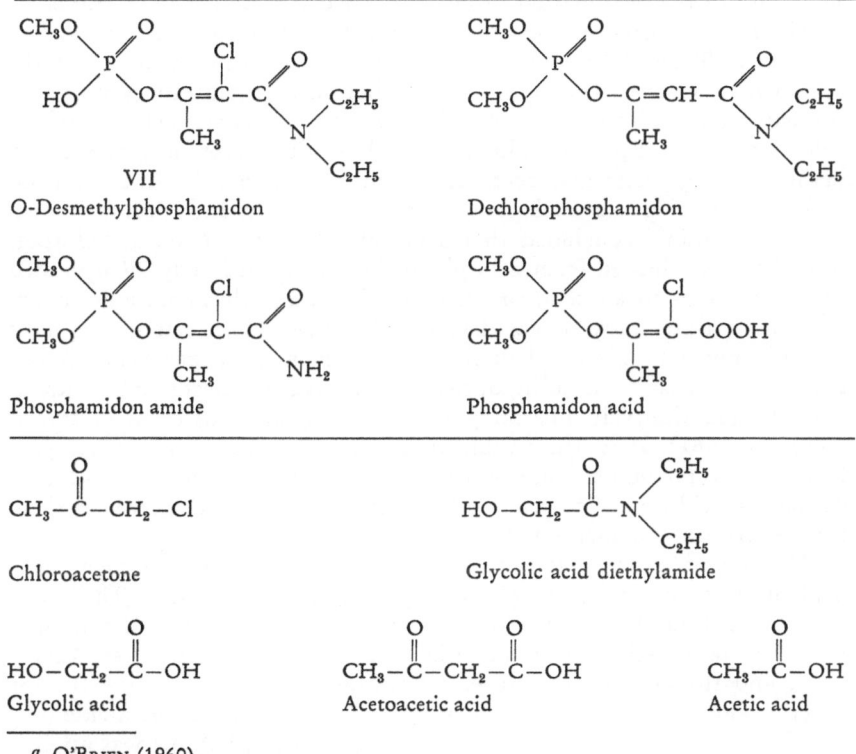

VII
O-Desmethylphosphamidon

Dechlorophosphamidon

Phosphamidon amide

Phosphamidon acid

Chloroacetone

Glycolic acid diethylamide

Glycolic acid

Acetoacetic acid

Acetic acid

[a] O'BRIEN (1960).
[b] Adapted from ANLIKER et al. (1961 b).

The absence of O-desmethyl phosphamidon (Table II), indicating lack of hydrolysis of the P-O-methyl linkage of phosphamidon, seemed surprising. Later investigation of the problem by a different experimental approach revealed that traces of O-desmethyl phosphamidon (VII) were in fact present in bean seedlings treated with the ^{14}C-labeled insecticide (*CIBA* 1964 b). In this study, advantage was taken of the esterification reaction of O-desmethyl phosphamidon with diazomethane to form the easily detectable parent compound *(CIBA* 1964 c, see also Chapter 1). When the aqueous residue obtained by the standard fractionation procedure (Fig. 2) was subjected to this esterification treatment, small but measureable amounts of phosphamidon were formed which were recovered by subsequent methylene chloride extraction. The quantities of O-desmethyl phosphamidon detected by this method four and eight days after insecticide application were less than one percent of the dose applied.

In a further search for the identity of the polar radioactive fraction mentioned above, Anliker *et al.* (1961 b) subjected phosphamidon to model hydrolysis experiments in aqueous acid and alkaline solutions and examined the nature of the degradation products formed (for details see Chapter 1).

The main hydrolytic breakdown products, which are listed in the bottom part of Table II, were then used as references to be compared with the unknown radioactive fraction. The distribution of radioactivity on paper chromatograms indicated that the unknown fraction contained none of the following potential phosphamidon metabolites: chloroacetone, glycolic acid diethylamide, glycolic acid, acetoacetic acid, and acetic acid. Anliker *et al.* (1961 b), pursuing the analogy with the alkaline hydrolytic breakdown of phosphamidon, concluded that some of the small dechlorinated fragments listed are indeed formed in plants, but are immediately incorporated into endogenous constituents of the plant tissue, and thus escape detection.

In a separate study it was demonstrated that chloroacetone was most probably not a persistent plant metabolite. In model experiments, it immediately reacted with sulfhydryl-containing tissue constituents, such as cysteine, quantitatively liberating the chloride ion *(California Chemical Company* 1961). It is conceivable that other chlorinated hydrolysis products of phosphamidon, such as the α-chloroacetoacetamides or α-chloroacetoacetic acid are dehalogenated by a similar mechanism; however, there is no experimental evidence for this.

The second major investigation on the metabolism of phosphamidon in plants was undertaken more recently by Bull *et al.* (1967). This work was started after the presence of the two geometrical isomers of phosphamidon in the technical grade material had been substantiated (see Chapter 1), and the authors, therefore, concentrated on following the fate of the two phosphamidon isomers separately. The ^{32}P-labeled insecticide (obtained from *Nuclear Chicago Corp.,* Des Plaines, Ill., U.S.A.) was applied by petiole injection to field-grown cotton plants and by foliar treatment to cotton seedlings grown in the greenhouse. In addition, the same ^{14}C-double labeled phosphamidon preparation which had been used by Anliker *et al.* (1961 b) and the ^{32}P-labeled insecticide were introduced into isolated cotton leaves and alfalfa sprigs by uptake from a nutrient solution.

At intervals of one to eight days after these various applications, Bull *et al.* (1967) extracted the plant material and fractionated the extracts according to procedures which had been described in detail for other vinyl-type insecticides (Bull and Lindquist 1964 and 1966). Their methods were essentially those of Anliker *et al.* (1961 b), and they took advantage of the particular partition coefficients of phosphamidon and its structural analogs which allowed removal of plant pigments from the aqueous starting extract by *n*-hexane without appreciable loss of radioactivity. The aqueous phase was then extracted with chloroform to recover the two isomers of the parent compound and their organosoluble transformation products. Radioactivity present in this organic extract was separated by reversed-phase paper chromatography and compared with authentic references. Separation of label remaining in the aqueous phase was likewise achieved on paper and cellulose thin-layer plates with a relatively polar solvent system (Fig. 4). Unfortunately, BULL *et al.* (1967) did not use any additional

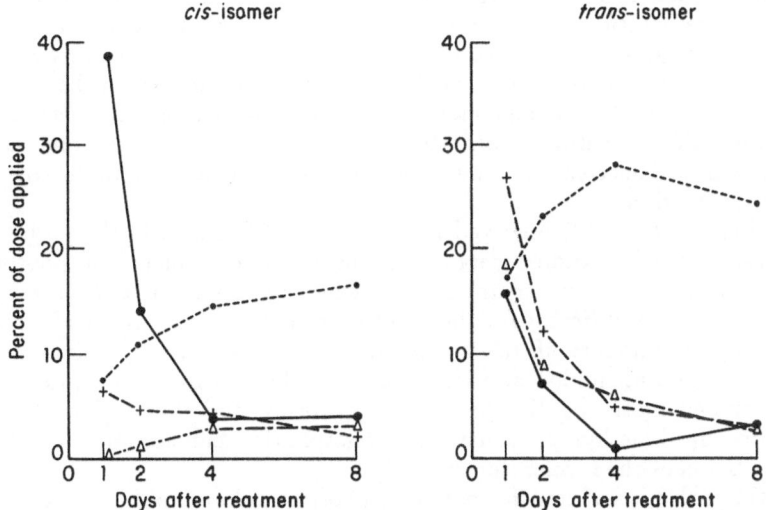

Fig. 4. Relative concentrations of ^{32}P-labeled phosphamidon (•—•) and its major metabolites, dimethylphosphate (•----•), O-desmethylphosphamidon (+--+), and N-desethylphosphamidon, in field-grown cotton leaves treated *in situ* by petiole injection (*cis*-isomer 111 µg./g. of leaf; *trans*-isomer 114 µg./g. of leaf). Curves prepared from data obtained by BULL *et al.* (1967); reproduced with permission of *J. Econ. Entomol.*

analytical means to characterize the separated metabolites. Identification of the observed transformation products by their partitioning behaviour and by one or two R_f values cannot, therefore, be considered as being more than tentative.

In all experiments with ^{32}P-labeled phosphamidon, dimethyl phosphate became the main metabolite within two days of the application (for a typical experiment see Fig. 4). This observation confirmed that hydrolysis of the P-O-vinyl linkage is the major route of phosphamidon degradation. Phosphoric acid was less abundant than dimethyl phosphate, whereas methyl phosphate was recovered only in traces. BULL *et al.* (1967) also confirmed the formation of N-desethyl-phosphamidon and O-desmethyl phosphamidon. In addition they suggested N-desethyl-O-desmethyl phosphamidon as a further transformation product. The amounts of the three last-mentioned metabolites relative to phosphamidon were somewhat higher than those detected by the CIBA scientists (Fig. 4). However, with the particular chromatographic systems used by Bull et al. (1967), O-desmethyl phosphamidon did not separate from N-desethyl-O-desmethyl phosphamidon and, consequently, neither the identities of these two metabolites nor their relative amounts were clearly established.

In following the fate of the non-phosphorus moiety of phosphamidon and N-desethyl phosphamidon with the ^{14}C-labeled compound, Bull et al. (1967) detected traces of radioactivity which co-chromatographed with α-chloroacetoacetic acid diethylamide and α-chloroacetoacetic acid ethyl-amide, respectively. Formation of these two primary hydrolytic metabolites as observed by Anliker et al. (1961 b) was thus confirmed. In the absence of suitable reference compounds, no further phosphorus-free metabolites were identified by Bull et al. (1967), although two minor and one major unknown radioactive spot were detected upon chromatography of the aqueous fraction.

Bull et al. (1967) observed no significant differences in the pathways of transformation and/or degradation of the two isomers[1] of phosphamidon in plants. However, the rates at which the different steps proceeded were apparently different for the two forms of the parent insecticide. From the relative amounts of the metabolites, it seemed that the biologically active cis-phosphamidon and its organo-soluble transformation product(s) decomposed more rapidly than the trans-forms, and that higher concentrations of dimethyl phosphate and phosphoric acid accumulated upon degradation of the former isomer.

The pathways of transformation of phosphamidon in plants, as revealed by the above investigations, are summarized in Figure 8 (see later). It is definitely established that hydrolysis of the P-O-vinyl linkage represents the main route of degradation. It seems that the primary phosphorus-free hydrolysis products then immediately decompose further, since none of the many potential metabolites examined were observed to accumulate in plant tissues. Hydrolysis of the P-O-methyl bond is the second pathway of degradation, but present data indicate that it is of minor importance. N-desethylation of phosphamidon produces the only biologically active metabolite which has so far been detected in plant material. As will be discussed later, there is at present no evidence for a second N-desethylation step, or for the formation of persistent intermediates or conjugates during oxidative N-dealkylation.

---

[1] In analogy to the cis- and trans-isomers of phosdrin the biologically active isomer of phosphamidon is that form in which the allylic methyl and diethylcarbamyl groups are in a cis-relationship (see Chapter 1). According to this convention, the nomenclature used by Bull et al. (1967) must be reconsidered (Spencer 1967). Their trans-isomer is in fact cis-phosphamidon and vice versa.

III. Metabolism in animals

To evaluate the toxicological significance of phosphamidon residues in treated commodities, we studied not only the metabolic pathways in the mammalian body, but also uptake, distribution, and elimination. Since a toxic metabolite (N-desethyl phosphamidon) occasionally occurs in small amounts on edible crops its fate in mammalian tissue had also to be considered (see preceding subchapter and Chapter 8).

Preliminary studies on the behaviour of phosphamidon in rats were conducted by the *CIBA*-laboratories (*CIBA* 1962 a). The same ^{14}C-double labeled phosphamidon as used in plant experiments (see preceding subchapter) was administered to several animals in a single dose of three mg./kg. Excretion of radioactivity in the urine was rapid, amounting to about 90 percent of the original amount within 72 hours. Smaller percentages were eliminated in the feces (three percent) and exhaled (two percent). An average of 95 percent had thus been eliminated by the end of the 72-hour experiment. On extraction and fractionation of the urine in accordance with a scheme similar to that used for plants (see Fig. 2), a very small percentage of label partitioned into methylene chloride (three percent of the total activity present in urine), and paper chromatography revealed that this organo-soluble radioactivity contained neither phosphamidon nor N-desethyl phosphamidon, both of which were consequently claimed to be below the 0.001 p.p.m. level in urine. Radioactivity remaining in the aqueous phase comprised at least three polar metabolites which were not further identified.

In the same series of experiments, tissue levels and the possibilities of accumulation of radioactivity in various organs were examined by feeding to groups of rats a diet containing one and ten p.p.m. of ^{14}C-labeled phosphamidon for five days. Two and four days, respectively, after feeding had been discontinued, the animals were sacrificed and the principal organs and tissues subjected to the standard extraction and fractionation procedure (Fig. 2). Total (methanol-) extractable radioactivity was observed to be not above, and usually much below, 0.1 percent of the dose applied in all organs and tissues examined, including lungs, liver, kidneys, brain, muscle, and fat. Radioactivity partitioning into methylene chloride, presumably representing phosphamidon, N-desethyl phosphamidon, and the substituted α-chloroacetoacetamides (see preceding subchapter), was negligible, and tissue levels of these compounds were calculated to be below 0.005 p.p.m.

To confirm the lack of accumulation of phosphamidon and/or persistent toxic metabolites in the animal body, two oxen were fed a diet containing 20 p.p.m. of the unlabeled insecticide for a period of five days (*CIBA* 1962 b). They were slaughtered one and five days after feeding of phosphamidon had been discontinued. Samples of muscle, fat, liver, kidneys, lungs, and brain were extracted, and the extracts cleaned up by the double-paper chromatography residue method (see Chapter 7). None of the developed paper chromatograms (for solvent system and color reaction see Fig. 1) revealed detectable amounts of phosphamidon, N-desethyl phosphamidon, or the substituted α-chloroacetoacetamides, which indicated that the concentration of the parent compound in these tissues was below 0.005 p.p.m. In a second feeding experiment, six lactating cows were fed for three days with grass which had been heavily sprayed with phosphamidon (0.04 percent active ingredient). Milk samples collected during and after the feeding period were examined for the presence of cholinesterase inhibitor(s). No inhibitor was detected in milk, but, unfortunately, neither the sensitivity of the method nor the appropriate recovery values was determined.

In parallel with the CIBA-experiments, Plapp and Eddy (1961) studied the fate of ¹⁴C-double labeled phosphamidon in white mice after subcutaneous and oral administration. Elimination of radioactivity, which was followed only in the urine, was again found to be rapid, and ceased within 72 hours of administration. About 50 percent of label was recovered in the urine, which was smaller than the percentage observed in the CIBA studies. When the animals were sacrificed three days after administration of the insecticide, no residual radioactivity was detected in tissue samples of brain, heart, lungs, liver, and kidney. With the particular isotope-counting system used this meant that total residues in these tissues were below 0.01 p.p.m. phosphamidon equivalents. When examining urine samples for metabolites by paper chromatography, the authors observed seven different radioactive peaks, but did not attempt to purify and identify these transformation products.

More recently, Bull et al. (1967) introduced ³²P-labeled isomers of phosphamidon into white rats by intraperitoneal injection, and kept them in metabolism cages for four days. As with the ¹⁴C-labeled compound, excretion of radioactivity was rapid accounting for 60 percent of the dose applied (for each of the two isomers) within the first 24 hours. Elimination was mainly by the urine, which contained only traces of the parent isomers of the insecticide. The principal hydrolysis product was dimethyl phosphate, with small concentrations of phosphoric acid and a radioactive compound which co-chromatographed with O-desmethyl phosphamidon.

Although the rapid degradation and elimination of phosphamidon and/ or its metabolites by the animal body and the apparent lack of accumulation of persistent transformation products in the tissues had been fairly well established by the experiments described, these had not revealed the pathways by which phosphamidon was actually degraded in mammals. In fact, none of the radioactivity excreted in the urine, which, admittedly, contained only small amounts of organosoluble substances, had been accounted for in terms of identified metabolites. Phosphamidon itself and the transformation products observed in plants (i. e., N-desethyl phosphamidon and the substituted α-chloroacetoacetamides) were apparently below detectable levels in urine, milk, and tissues of mammals.

In an effort to obtain a clearer picture of the degradation of phosphamidon in mammalian tissues, several "in vitro" and short term "in vivo" experiments were conducted. Rat liver homogenates fortified with appropriate co-factors (magnesium chloride, nicotinamide, NADP) were found to transform and degrade ^{14}C-labeled phosphamidon (*CIBA* 1964 d). After 90 minutes of incubation, 63 percent of the radioactivity was recovered in methylene chloride upon extraction and fractionation of the mixture by the standard method (Fig. 2). Paper chromatography of the organic extract revealed the presence of three radioactive compounds, which co-chromatographed with phosphamidon (*cis*- and *trans*-mixture) and *cis*-

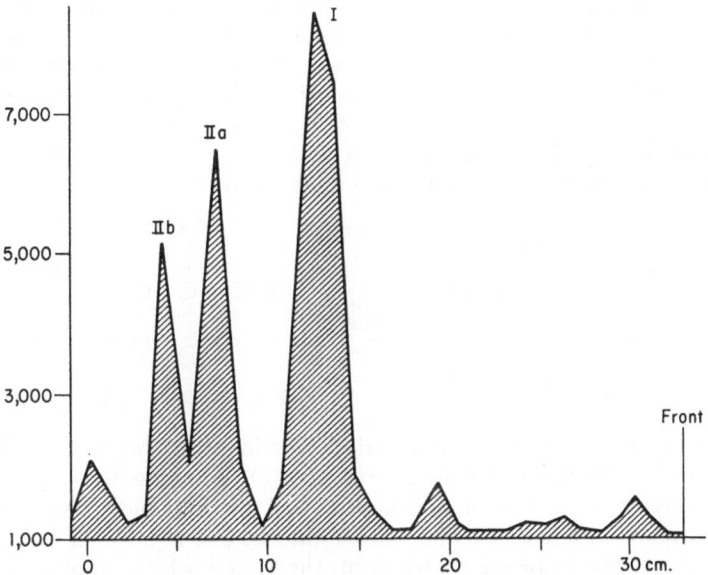

Fig. 5. Separation by reversed-phase paper chromatography of radioactivity recovered in methylene chloride extract derived from rat liver homogenate incubated with ^{14}C-double labeled phosphamidon. Reaction stopped 90 minutes after phosphamidon addition. Solvent system B$_1$ according to Bush (1952; see Fig. 3). I = phosphamidon (*cis*- and *trans*-mixture), II *a* = *cis*-*N*-desethyl phosphamidon, and II *b* = *trans*-*N*-desethyl phosphamidon (*CIBA* 1964 d)

N-desethyl phosphamidon and *trans*-*N*-desethyl phosphamidon, respectively (Fig. 5). *N*-dealkylation of phosphamidon was thus demonstrated for animal tissues.

The occurrence of the *N*-desethylation reaction was confirmed in living animals *(Industrial Bio-Test Laboratories* 1965 a). White rats received small quantities of phosphamidon by intubation, and were killed five minutes later. Their livers, which were analyzed according to the double-paper chromatography residue method (see Chapter 7), contained small but measurable quantities of *N*-desethyl phosphamidon in addition to the unchanged insecticide. The substituted α-chloroacetoacetamides were not detected. The rapid appearance of phosphamidon in liver observed in these experiments demonstrated that the insecticide is rapidly absorbed from the gastrointestinal tract.

In the above short-term experiments, the primary phosphorus-free hydrolysis products of phosphamidon and N-desethyl phosphamidon, present in traces in plants (see preceding subchapter), had not been detected as yet in animal tissues. Model experiments demonstrated that these substances were apparently rapidly dechlorinated by a purely hydrolytic process (*Industrial Bio-Test Laboratories* 1965 b). When α-chloroacetoacetic acid diethylamide was kept in slightly alkaline aqueous solution (bicarbonate buffer pH 9.0) which would correspond to the conditions in the small intestine of mammals, the substance was observed to disappear rapidly, with parallel liberation of inorganic chloride. The dechlorinated hydrolysis product, purified and isolated as the 2,4-dinitrophenylhydrazine-derivative, was identified by elementary analysis and comparison with an authentic standard as α-hydroxyacetoacetic acid diethylamide:

$$CH_3-\overset{\overset{\displaystyle O}{\|}}{C}-\underset{\underset{\displaystyle Cl}{|}}{CH}-C\overset{\diagup O}{\underset{\diagdown N\diagdown C_2H_5}{\diagdown C_2H_5}} \quad \xrightarrow[-Cl^-]{OH^-} \quad CH_3-\overset{\overset{\displaystyle O}{\|}}{C}-\underset{\underset{\displaystyle OH}{|}}{CH}-C\overset{\diagup O}{\underset{\diagdown N\diagdown C_2H_5}{\diagdown C_2H_5}}$$

$$[III] \qquad\qquad\qquad\qquad\qquad\qquad [V]$$

The same hydrolytic dechlorination reaction was shown to occur *"in vivo"* (*Industrial Bio-Test Laboratories* 1965 c). When ligated dog intestine was intubated with α-chloroacetoacetic acid diethylamide, the substance disappeared with a half-life of 15 to 20 minutes (see Fig. 6). No substituted amide was found in blood drawn from the vessels which supplied the segment of the intestinal loop.

Up to this stage, animal metabolism experiments with phosphamidon were guided by the assumption that pathways of transformation were similar to or identical with those observed in plants. The mechanisms of hydrolysis of the P-O-vinyl linkage and N-desethylation had indeed been demonstrated also in mammalian tissues and these mechanisms apparently accounted for the transformation of the bulk of the insecticide in various animal species. No other pathways had been discovered in mammals.

Evidence for animal metabolites of phosphamidon, which apparently were not to be found in plants, was presented by Clemons and Menzer (1968) following a thorough and detailed study of the metabolism of the insecticide in mammalian tissues. Two different radioactive preparations were used, one labeled with ^{32}P, the other labeled with ^{14}C in the N-diethyl moiety. The insecticide was administered orally to white rats and to a lactating goat. Excretion of radioactivity was followed in the urine of both species and in the milk from the goat. Urine and milk samples were fractionated to allow characterization of the organo-(chloroform-)soluble

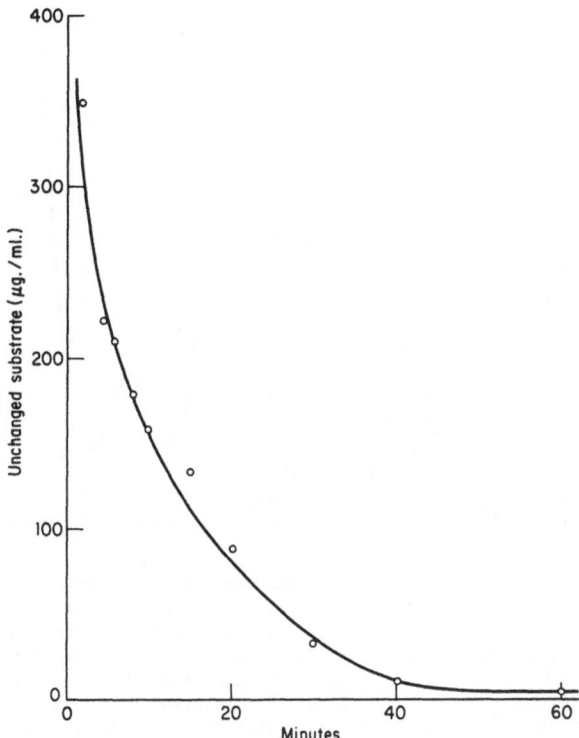

Fig. 6. Rate of hydrolytic dehalogenation of α-chloroacetoacetic acid diethylamide by the ligated dog intestine. Concentrations of substrate determined by gas chromatography using flame ionisation detection (*Industrial Bio Test Laboratories* 1965 c)

metabolites. These metabolites were separated on a liquid/liquid partitioning Celite column (see Fig. 7) and by paper chromatography with solvent systems described by ANLIKER and MENZER (1963) and BULL *et al.* (1967). Further characterization of the metabolites was based on determination of their ^{32}P-to-^{14}C ratios (Fig. 7).

CLEMONS and MENZER (1968) observed rapid excretion of radioactivity in rat and goat urine, confirming the results of earlier experiments. About 70 percent of the administered ^{32}P-labeled dose was eliminated within the 72-hour sampling period. Excretion of ^{14}C in urine was less complete, amounting to approximately 50 percent. Elimination of ^{14}C in the feces and by exhalation, which might be substantial with an *N*-diethyl-labeled compound, was unfortunately not examined. Over 90 percent of the radioactivity recovered in the urine samples represented water-soluble products of hydrolysis, which were assumed to correspond to the non-toxic compounds and fragments detected in earlier plant experiments (see preceding subchapter) and which were not further analyzed.

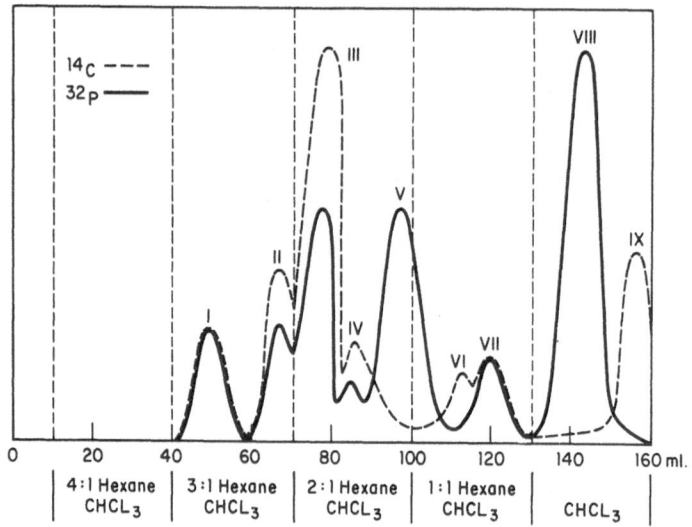

Fig. 7. Chromatographic characteristics of phosphamidon metabolites on Celite column based on partitioning between water and hexane-chloroform mixtures. Reproduction of results obtained from chromatography of chloroform-extractable material from one-hour sample of goat urine. I = desethylphosphamidon; II, IV, VI and VII = minor unknowns, III = major unknown, V = phosphamidon amide, VIII = dechlorophosphamidon amide, IX = α-chloro-N,N-diethyl acetoacetamide. Adapted from Clemons and Menzer (1968) and reproduced with permission of *J. Agr. Food Chem.*

A total of nine different organosoluble metabolites was detected in either the urine of both animal species or in the goat milk. Of these, four were structurally identified by co-chromatography on column and paper with authentic samples and by their isotope ratio. In addition to N-desethyl phosphamidon and the phosphorus-free hydrolysis product α-chloroacetoacetic acid diethylamide, the N-unsubstituted amide of phosphamidon (VIII) and the dechlorinated, unsubstitued amide (IX) were recognized:

[VIII]

[IX]

CLEMONS and MENZER (1968) had thus shown that phosphamidon is subject to a second oxidative dealkylation step and to reductive dehalogenation in animal tissues. The products of these reactions are strong anticholinesterases which are apparently somewhat more toxic than the parent compound. However, their quantities in urine were very small, amounting to no more than 0.1 to 0.01 percent or less of the dose applied. In goat milk, their concentration was of the order of 0.01 p.p.m., although the oral dose administered was very high (three mg./kg.) in terms of practical insecticide applications.

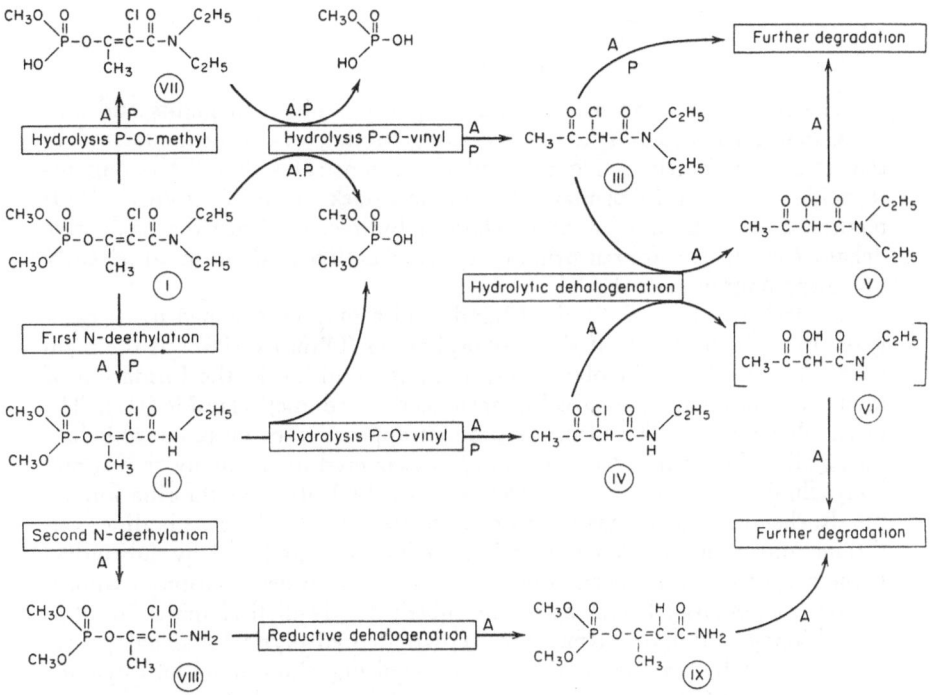

Fig. 8. Pathways of degradation of phosphamidon in animals (A) and plants (P)

Among the unidentified metabolites of CLEMONS and MENZER (1968) two were present in very minute amounts, whereas a third contained only ^{14}C and no ^{32}P and was consequently assumed to be a hydrolysis product. An additional metabolite, somewhat more abundant than the identified substances, exhibited the original ^{32}P-to-^{14}C ratio of phosphamidon, but did not correspond to either dechlorophosphamidon or the hydroxyethyl derivatives of phosphamidon and N-desethyl phosphamidon, respectively. Unfortunately, CLEMONS and MENZER (1968) did not take into account the possibility of isomer-separation of phosphamidon metabolites, which our experience at *CIBA* has shown to be possible under specific conditions.

The animal metabolism experiments described in detail show that phosphamidon is rapidly degraded in and eliminated from the mammalian body. The main pathway of degradation is hydrolytic, involving cleavage of the P-O-vinyl bond to form dimethyl phosphate and small dechlorinated fragments (see Fig. 8). Minor pathways include stepwise oxidative *N*-dealkylation and reductive dehalogenation, producing traces of toxic metabolites which are apparently also rapidly hydrolyzed, the products being eliminated. There is at present no evidence for the formation of persistent toxic metabolites.

IV. Discussion

The pathways of transformation of phosphamidon in plants and animals, described in detail in the preceding subchapters, are summarized in the scheme presented in Figure 8, which demonstrates that the parent insecticide is subject to primary enzymatic attack at the P-O-vinyl, P-O-methyl, and *N*-diethyl linkages. There is indirect evidence that reductive dehalogenation is a fourth primary transformation mechanism, apparently operating only in animals.

Hydrolysis of the P-O-vinyl bond, which may be assumed to be catalyzed by different types of tissue phosphatases (O'Brien 1967), is the main route of degradation in plants and animals, resulting in the formation of dimethyl phosphate and α-chloroacetoacetic acid diethylamide (III). The latter phosphorus-free hydrolysis product appears to be quite unstable biochemically, since only traces have been discovered in the many tissues and body fluids examined for its presence. Its hydrolytic dechlorination by purely chemical means has been demonstrated under slightly alkaline conditions and a similar dehalogenation mechanism, catalyzed by sulfhydryl tissue constituents, has been observed to cause rapid degradation of chloroacetone, a potential fragment of the substituted *N*-diethylamide. In addition, kidney and liver tissues are known to hydrolytically dehalogenate alkyl halides by an enzymatic process requiring glutathione and cyanide ions (Parke 1968). It may therefore be safely assumed that no persistent chlorinated metabolites derived from phosphamidon accumulate in plant and animal tissues. The dechlorinated hydrolytic degradation products (V, VI) are subject to further cleavage by deamidation and decarboxylation, producing small fragments which enter the common carbon metabolite pool of the organism.

Hydrolysis of the P-O-methyl bond, leading to the formation of the non-toxic, polar metabolite O-desmethyl phosphamidon (VII), is a minor pathway of degradation in plants and animals. The quantities of the metabolite and of its methyl phosphate hydrolysis product which have been detected in various tissues are rather small. However, it has to be remembered that methyl phosphate, unlike dimethyl phosphate, is rapidly hydrolized to phosphoric acid (O'Brien 1967), and this conversion may mask the actual quantities of methyl phosphate being formed in tissues.

Oxidative N-dealkylation is the only pathway operative in plants and animals which leads to the formation of toxic metabolites of phosphamidon. Whereas plants seem to be able to perform the first dealkylation step only, animals can apparently remove the second N-ethyl group as well. Oxidative N-dealkylation which has been studied most extensively with liver microsomal systems, requires molecular oxygen and reduced NADP. There is increasing evidence that the mechanism of N-dealkylation involves the enzymic formation of an N-oxide and/or a hydroxyalkyl intermediate, which subseqently undergoes enzymic or spontaneous molecular rearrangement to yield the unsubstituted amine or amide (VIII) plus an aldehyde (Gilette 1963, McMahon 1964, Parke 1968).

Dicrotophos (Bidrin®, Carbicron®) and monocrotophos (Azodrin®, Nuvacron®), two non-chlorinated vinyl-phosphate insecticides whose structure is quite similar to that of phosphamidon, have been shown to be N-dealkylated with the intermediate formation of the corresponding N-hydroxymethyl (or methylol-) derivatives (Bull and Lindquist 1964 and 1966, Menzer and Casida 1965). As for phosphamidon itself, Bull et al. (1967) have claimed that a radioactive peak with chromatographic properties corresponding to those of N-hydroxy ethyl phosphamidon is indeed present in small quantities in cotton plants, boll worms, and boll weevils. Unfortunately, no details were given on the synthesis and stability of the reference compound which was used in these experiments. Clemons and Menzer (1968) also attempted to demonstrate the presence of N-hydroxyethyl derivates of phosphamidon in urine, by liberating from unidentified metabolites the corresponding acetaldehyde and trapping the latter compound as the semicarbazone. Clemons and Menzer (1968) were unable to prove the existence of N-hydroxyethyl intermediates of phosphamidon by this procedure. Several attempts in the CIBA-laboratories to synthesize the N-hydroxyethyl derivatives of phosphamidon and of N-desethyl phosphamidon failed owing to the pronounced instability of the compounds. It may, therefore, be concluded that if such oxidative intermediates of phosphamidon are formed under biological conditions they will probably decompose rapidly.

The N-hydroxymethyl intermediates of dicrotophos and monocrotophos form relatively persistent carbohydrate conjugates in certain plant tissues (Bull and Lindquist 1964 and 1966), and these have to be taken into account in analysis of plant commodities for these insecticides. So far there are no indications that N-hydroxyethyl derivatives or other toxic metabolites of phosphamidon exhibit similar conjugation (Bull et al. 1967). The N-hydroxyethyl intermediates of phosphamidon appear to be too unstable to be trapped as conjugates.

Reductive dechlorination of phosphamidon is suggested from the appearance of small quantities of dechlorophosphamidon amide (IX) in rat urine and goat milk (Clemons and Menzer 1968). Although reductive

dehalogenation has so far not been recognized as a widespread mechanism
of pesticide transformation, certain drugs have been observed to be so
metabolized by a hepatic microsomal enzyme which requires oxygen and
reduced NADP (van Dyke and Chenoweth 1965).

No attempt has been made in this chapter to discuss the toxicological
implications of the formation of the limited number of toxic or potentially
toxic animal and plant metabolites of phosphamidon. These pertinent experiments and results are described in detail in Chapter 5.

Summary

Enzymatic degradation of phosphamidon occurs at different sites of
the molecule. The major route of breakdown is P-O-vinyl hydrolysis,
resulting in the formation of dimethylphosphate and α-chloro-acetoacetic
acid diethylamide. The latter compound is probably dehalogenated and
further cleaved by deamination and decarboxylation. P-O-methyl hydrolysis of phosphamidon does also occur in plants and animals, but it is a
minor pathway of degradation. The resulting metabolite, O-demethyl-
phosphamidon, is non-toxic. The third mechanism, oxidative N-desethylation, leads to the formation of N-desethyl-phosphamidon (plants and animals), and phosphamidon amide (animals). Both metabolites are toxic. The
N-hydroxyethyl intermediates, however, and their possible carbohydrate
conjugates have not been detected so far. Finally reductive dechlorination
is suggested from the appearance of small quantities of dechlorophosphamidon amide in rat urine and goat milk.

Résumé *

Le métabolisme du phosphamidon chez les plantes et les animaux

Les transformations enzymatiques du phosphamidon atteignent différentes fonctions de la molécule. La voie majeure de dégradation est
représentée par l'hydrolyse de la liaison P-O-vinyle. Cette hydrolyse mène
au diméthylphosphate et à l'acide α-chloro-acétoacétique diéthylamide. Ce
dernier composé est probablement déhalogéné et ensuite clivé par desami-
nation et décarboxylation. L'hydrolyse de la liaison P-O-méthyle du phosphamidon se produit aussi bien chez les plantes que chez les animaux, mais
c'est une voie de dégradation mineure. Le métabolite obtenu: O-déméthyl-
phosphamidon, n'est pas toxique. Le troisième mécanisme de transformation, la N-deséthylation oxydative, conduit à la formation de N-deséthylphosphamidon (plantes et animaux) et de l'amide du phosphamidon (animaux). Les deux métabolites sont toxiques. Toutefois les dérivés N-hydroxy
éthyles n'ont pas été mis en évidence comme intermédiaires. Il en est de
même pour les conjugés glucosidiques de ces dernières substances. Finalement, la détection de faibles quantités de deschlorophosphamidon amide
dans l'urine de rat et le lait de chèvre suggère une perte réductive de chlore.

---

* Traduit par les auteurs.

Zusammenfassung *

Der Stoffwechsel des Phosphamidons in Pflanzen und Tieren

Phosphamidon wird an verschiedenen Stellen des Moleküls enzymatisch abgebaut. Im Vordergrund steht die P-O-Vinyl-Hydrolyse, die zur Bildung von Dimethylphosphat und α-Chloracetessigsäurediäthylamid führt. Die letzterwähnte Substanz wird möglicherweise dehalogeniert und durch Deaminierung und Decarboxylierung weiter gespalten. Eine P-O-Methyl-Hydrolyse findet in Pflanzen und Tieren ebenfalls statt, allerdings nur in sehr beschränktem Ausmaß. Der durch diesen Abbauweg gebildete Metabolit Demethylphosphamidon ist ungiftig. Der dritte Mechanismus wird durch die oxidative N-Deäthylierung repräsentiert. Dieser Vorgang führt zur Bildung von N-Deäthylphosphamidon (Pflanzen und Tiere) und Phosphamidon-Amid (Tiere). Beide Metaboliten sind toxisch. Der Nachweis von N-Hydroxyäthylphosphamidon, mit großer Wahrscheinlichkeit eine Zwischenstufe bei der N-Deäthylierung, gelang bisher nicht; dasselbe gilt für die Glykoside dieser Substanz. Da in Rattenurin und Ziegenmilch Spuren von Deschlorphosphamidon nachgewiesen wurden, erscheint auch eine reduktive Abspaltung von Chlor möglich zu sein.

References

ANLIKER, R.: Metabolism of the systemic insecticide phosphamidon. Presented Vth Internat. Pesticides Congress, London (1963).

—, E. BERIGER, und K. SCHMID: Die Synthese von ¹⁴C-markiertem Phosphamidon, einem neuen systemischen Insektizid. Experientia 17, 492 (1961 a).

— —, M. GEIGER, und K. SCHMID: Über die Synthese von Phosphamidon und seinem Abbau in Pflanzen. Helv. Chim. Acta. 44, 1622 (1961 b).

—, and R. E. MENZER: Method for phosphamidon residue analysis. J. Agr. Food. Chem. 11, 391 (1963).

BULL, D. L., and D. A. LINDQUIST: Metabolism of 3-hydroxy-N,N-dimethylcrotonamide dimethyl phosphate by cotton plants, insects and rats. J. Agr. Food Chem. 12, 310 (1964).

— Metabolism of 3-hydroxy-N-methyl-cis-crotonamide dimethyl phosphate (Azodrin) by insects and rats. J. Agr. Food Chem. 14, 105 (1966).

—, and R. R. GRABBE: Comparative fate of the geometric isomers of phosphamidon in plants and animals. J. Econ. Entomol. 60, 332 (1967).

BUSH, I. E.: Methods of paper chromatography of steroids applicable to the study of steroids in mammalian blood and tissues. Biochem. J. 50, 370 (1952).

California Chemical Co. (now Chevron Chemical Co.): Phosphamidon decomposition products: Reactions of chloroacetone. Unpublished report (1961).

CASIDA, J. E.: Metabolism of organophosphate insecticides by plants: A review. In "Radioisotopes and radiation in entomology". Internat. Atomic Energy Agency, Vienna (1962).

CIBA Ltd.: Investigations on the excretion of phosphamidon by rats given single doses or diets containing ¹⁴C-labelled phosphamidon. Unpublished report (1962 a).

— Phosphamidon in tissues and organs of two oxen given a diet containing 20 p.p.m. of active compound. Unpublished report (1962 b).

— Isolation and identification of α- and β-isomers of desethyl-phosphamidon. Unpublished report (1964 a).

---

* Übersetzt von den Autoren.

CIBA Ltd.: Metabolism of phosphamidon in plants. Identification of desmethylphosphamidon in plant extracts. Unpublished report (1964 b).
— Method for desmethylphosphamidon residue analysis. Unpublished report (1964 c).
— Metabolism of phosphamidon by rat liver homogenates. Unpublished report (1964 d).
— Analysis of phosphamidon residues. Revised ed. Unpublished report (1964 e).
CLEMONS, G. P., and R. E. MENZER: Oxidative metabolism of phosphamidon in rats and a goat. J. Agr. Food Chem. **16**, 312 (1968).
FUKUTO, T. R., and R. L. METCALF: Metabolism of insecticides in plants and animals. Ann. N.Y. Acad Sci. **160**, 97 (1969).
GILLETTE, J. R.: Metabolism of drugs and other foreign compounds by enzymatic mechanisms. Fortschr. Arzneimitt. Forsch. **6**, 13 (1963).
HEATH, D. F.: Organophosphorus poisons. New York: Pergamon Press (1961).
Industrial Bio-Test Laboratories, Inc: In "vivo" metabolic conversion of phosphamidon to desethylphosphamidon. Report to *California Chemical Co.* (1965 a).
— Hydrolysis of *N,N*-diethyl-α-chloroacetoacetamide in mild alkaline medium. Report to *California Chemial Co.* (1965 c).
— In vivo hydrolysis of *N,N*-diethyl-α-chloroacetoacetamide in intact dog intestine. Report to *California Chemical Co.* (1965 c).
JAQUES, R., and H. J. BEIN: Toxikologie und Pharmakologie eines neuen systemisch wirksamen Insektizids der Phosphorsäureester-Reihe, Phosphamidon (2-Chlor-2-diäthylcarbamoyl-1-methylvinyl-dimethylphosphat). Arch. Toxikol. **18**, 316 (1960).
MCMAHON, R. E.: Microsomal dealkylation of drugs. Substrate specificity and mechanism. J. pharm. Sci. **55**, 457 (1966).
MENZER, R. E., and J. E. CASIDA: Nature of toxic metabolites formed in mammals, insects and plants from 3-(dimethoxy-phosphinyloxy)-*N,N*-dimethyl-cis-crotonamide and its *N*-diethyl analog. J. Agr. Food Chem. **13**, 102 (1965).
O'BRIEN, R. D.: Toxic phosphorus esters. New York: Academic Press (1960).
— Insecticides. Action and metabolism. New York: Academic Press (1967).
PARKE, D. V.: The biochemistry of foreign compounds. Oxford: Pergamon Press (1968).
PLAPP, F. W., and G. W. EDDY: Studies on the metabolism of ^{14}C-labelled phosphamidon. Unpublished report U.S. Agr. Research Service, *U.S. Department of Agriculture* to *California Chemical Company* (1961).
SPENCER, E. Y.: Nomenclature of phosphamidon isomers. J. Econ. Entomol. **60**, 1749 (1967).
—, R. D. O'BRIEN, and R. W. WHITE: Permanganate oxydation products of schradan. J. Agr. Food Chem. **5**, 123 (1957).
TSUYUKI, H., M. A. STAHMANN, and J. E. CASIDA: Preparation, purification, isomerization and biological properties of octamethyl-pyrophosphoramide *N*-oxide. J. Agr. Food Chem. **3**, 922 (1955).
VAN DYKE, R. A., and M. B. CHENOWETH: The metabolism of volatile anaesthetics. II. In vitro metabolism of methoxyflurane and halothane in rat liver slices and cell fractions. Biochem. Pharmacol. **14**, 603 (1965).
WILLIAMS, C. T.: Detoxification mechanisms. London: Chapman & Hall (1949).

Chapter 5

Toxicology of phosphamidon

-By

K. R. Sachsse and G. Voss

Contents

I. Introduction

The toxicity of phosphamidon and of its main metabolites, by-products, and formulations has been investigated by means of acute, subacute, and chronic toxicity tests on various vertebrate species in many parts of the world. Particular attention has been paid to acute oral, dermal, and inhalation toxicity testing in mammals, and to the effects of repeated dosing over long periods. Enzymatic and histopathologic investigations were also needed to clarify the mode of action. Special trials covering skin irritation, potentiation, and neurotoxicity provided additional information with regard to public health.

The toxicity of phosphamidon to various birds, fish, and aquatic insects has also been studied in several field trials and during the course of large-scale aerial spraying campaigns.

In this article we review the results of this work to date. The increasing wide-scale use of phosphamidon will necessitate further toxicity work, with particular reference to public health and contamination of the environment.

Table I. *Determination of acute oral toxicity of technical grade phosphamidon and its isomers*

Species and sex	Material	Vehicle	Acute oral LD_{50} (mg./kg. bodyweight)	References
Mouse, male	Phosphamidon, tech.	Water	11.2	*National Hygienic Laboratory* (1958)
Mouse, female	Phosphamidon, tech.	Water	9.0 ± 1.0	*Industrial Bio-Test Laboratories* (1964 d)
Rat, male	Phosphamidon, tech.	Propylene-glycol	30	Bᴜᴄʜᴇʀ (1959)
Rat, male	Phosphamidon, tech.	Corn oil	28.3 (24.4–32.8)	*Industrial Bio-Test Laboratories* (1961 b)
Rat, male and female	Phosphamidon, tech.	Water	17.2 (12.7–23.2)	*Biologisch-Medizinische Forschungsstation* (1969 e)
Rabbit, male	Phosphamidon, tech.	Propylene-glycol	70 (LD_{100})	Jᴀǫᴜᴇs and Bᴇɪɴ (1960)
Dog, male and female	Phosphamidon, tech.	Gelatine capsules	50	*Biologisch-Medizinische Forschungsstation* (1969 f)
Mouse, male and female	*Cis*-phosphamidon	0.9% NaCl solution	6.5	*CIBA*
Mouse, male and female	*Trans*-phosphamidon	0.9% NaCl solution	220	*CIBA*

Table II. *Determination of acute oral toxicity of phosphamidon metabolites*

Species and sex	Material	Vehicle	Acute oral LD$_{50}$ (mg./kg. bodyweight)	References
Mouse, male	α-Chloroaceto-acetdiethylamide	Propyleneglycol	> 1,000	JAQUES and BEIN (1960)
Rat, male	α-Chloroaceto-acetdiethylamide	Propyleneglycol	> 3,000	JAQUES and BEIN (1960)
Mouse, male	Desethylphosphamidon	Propyleneglycol	15	JAQUES and BEIN (1960)
Rat, male	Desethylphosphamidon	Propyleneglycol	25	JAQUES and BEIN (1960)
Rat, male	Desethylphosphamidon	Corn oil	37.2 (32.8–42.2)	*Industrial Bio-Test Laboratories* (1963)
Rat, male	Desethylphosphamidon	Corn oil	28.5 (20.6–38.3)	*Industrial Bio-Test Laboratories* (1966 d)
Rat	*Cis*-desethyl-phosphamidon	—	8.5	*CIBA*
Rat	*Trans*-desethyl-phosphamidon	—	250	*CIBA*
Mouse, male	α-Chloroaceto-acetethylamide	Propyleneglycol	750	JAQUES and BEIN (1960)
Rat, male	α-Chloroaceto-acetethylamide	Propyleneglycol	700	JAQUES and BEIN (1960)
Rat, male	Desmethylphosphamidon	Water	2,500 ± 420	*CIBA* (1964 a)

Table III. *Determination of acute oral toxicity with components of impurities present in technical grade phosphamidon and of formulated material (commercial products)*

Species and sex	Material	Vehicle	Acute oral LD$_{50}$ (mg. active substance/ kg. bodyweight)	References
Mouse, female	γ-Chlorophosphamidon	Water	117 ± 25	*Industrial Bio-Test Laboratories* (1964 d)
Rat, female	γ-Chlorophosphamidon	Water	126 ± 6.8	*Industrial Bio-Test Laboratories* (1964 d)
Rat, male	Dechlorophosphamidon	Corn oil	600 ± 70.1	*Industrial Bio-Test Laboratories* (1965 b)
Rat, male and female	Dimecron® 20	Water	14.0 (11.3–17.4)	*Biologisch-Medizinische Forschungsstation* (1969 c)
Rat, female	Dimecron 20	Water	15	Edson et al. (1965)
Rat, male and female	Dimecron 50	Water	30.0 (24.2–39.7)	*Biologisch-Medizinische Forschungsstation* (1969 c)
Rat, male and female	Dimecron 100	Water	12.0 (9.6–14.9)	*Biologisch-Medizinische Forschungsstation* (1969 c)

Table IV. *Determination of acute intraperitoneal (i. p.), subcutaneous (s. c.), and intravenous (i. v.) toxicity of technical grade phosphamidon and desethylphosphamidon*

Species and sex	Route	Material	Vehicle	Acute LD_{50} (mg./kg. bodyweight)	References
Mouse, male	i.p.	Phosphamidon, tech.	Water	6	CLEMONS and MENZER (1968)
Mouse, male and female	i.p.	Phosphamidon, tech.	0.85% NaCl solution	7.6 ± 0.8	*Industrial Bio-Test Laboratories* (1960 c)
Mouse, male and female	i.p.	Phosphamidon, tech.	Water	8.16 (7.22–9.23)	*Biologisch-Medizinische Forschungsstation* (1969 b)
Mouse, male	i.p.	Desethylphosphamidon	Water	7	CLEMONS and MENZER (1968)
Mouse, male	i.v.	Phosphamidon, tech.	Propyleneglycol and distilled water 1:9	6	JAQUES and BEIN (1960)
Mouse, male and female	i.v.	Phosphamidon, tech.	0.85% NaCl solution	5.6 ± 0.3	*Industrial Bio-Test Laboratories* (1960 c)
Rabbit, male and female	i.v.	Phosphamidon, tech.	Propyleneglycol	30 (LD_{100})	JAQUES and BEIN (1960)
Rat, male	s.c.	Phosphamidon, tech.	Propyleneglycol	26	JAQUES and BEIN (1960)

Table V. *Determination of acute percutaneous toxicity of technical grade phosphamidon and formulations (undiluted)*

Species and sex	Material	Method or reference	Acute percutaneous LD$_{50}$ (mg. active substance/kg.)	References
Rat, male	Phosphamidon, tech.	Abdominal application	530	Klotzsche (1958)
Rat, male and female	Phosphamidon, tech.	Noakes and Sanderson (1969)	374 (247–564)	Biologisch-Medizinische Forschungsstation (1969 a)
Rat, male and female	Dimecron 20	—	125	Edson et al. (1965)
Rat, male	Dimecron 20	Abdominal application	640	Klotzsche (1958)
Rat, male	Dimecron 20	Solution dropped onto skin	370	Klotzsche (1964)
Rat, male	Dimecron 20	Gauze plus tight dressing	460	Klotzsche (1964)
Rat, male	Dimecron 20	Gauze soaked with compound	530	Klotzsche (1964)
Rat, male and female	Dimecron 20	Noakes and Sanderson (1969)	260 (146–464)	Biologisch-Medizinische Forschungsstation (1969 d)
Rat, male and female	Dimecron 50	Noakes and Sanderson (1969)	250 (127–485)	Biologisch-Medizinische Forschungsstation (1969 d)
Rat, male and female	Dimecron 100	Noakes and Sanderson (1969)	250 (164–380)	Biologisch-Medizinische Forschungsstation (1969 d)
Rabbit, male and female	Phosphamidon, tech.	Shaved back	267 ± 38	Industrial Bio-Test Laboratories (1960 c)

II. Toxicity to laboratory animals

The following chapter deals with results of acute, subacute, and chronic toxicity studies on a range of laboratory animals, to compare the reactions of the different species. A two-year study in rats, virtually covering the life span of this animal, is important from the standpoint of possible carcinogenicity and accumulation. Long term studies provide a "no effect level" which is especially useful when calculating the "acceptable daily intake" for man.

a) *Acute toxicity*

To obtain an initial impression of the resorption of phosphamidon by the intestine, peritoneum, mucous membrane, and skin, acute toxicity tests were performed on adult mice, rats, rabbits, and dogs with the technical grade substance and its isomers, main metabolites, by-products, and formulations by various routes (oral, intraperitoneal, percutaneous, subcutaneous, intravenous, and respiratory); see Tables I to V for details.

The symptoms observed, typical for organophosphates, were salivation, lachrymation (mixed with secretion from the Harderian glands), exophthalmus, tachypnoea, hyperactivity, ataxia, tremors, generalized muscle spasms, clonic-tonic convulsions, prostration, and, in most cases, diarrhea. Time of onset of clinical symptoms, death, and recovery period of survivors after acute intraperitoneal, oral and percutaneous phosphamidon application to mice, rats and dogs are given in Table VI. Autopsy of the animals which died during the acute studies, and of those killed after the observation period, revealed no gross pathological changes attributable to phosphamidon.

Table VI. *Time schedule for clinical symptoms, death, and recovery of survivors after acute phosphamidon poisoning*

Route of application	Animal species	Time of onset of symptoms	Death within	Recovery of survivors
Intraperitoneal	Mouse	3 min.	5 min.– 3 hrs.	24 hrs.
Oral	Rat and dog	5 min.	20 min.–20 hrs.	48 hrs.
Percutaneous	Rat	3–5 hrs.	12–24 hrs.	10 days

Acute inhalation studies of particular interest in connection with possible hazard to spray operators and workers engaged in manufacture were carried out with male albino rats to determine the effects of phosphamidon and Dimecron 20 (eight g. active ingredient/cu. m. of air/three hrs.) on the respiratory organs. This dosage did not produce visible symptoms of poisoning (KLOTZSCHE 1958).

The LC_{50} of technical grade phosphamidon, calculated from a four-hour aerosol inhalation study in male and female rats (bodyweight 250 to

Table VII. Results of 90- and 120-day oral toxicity studies with technical grade phosphamidon on rats and dogs [a]

Animals male/female	Dosage levels	Behaviour of cholinesterases [b]						Pathol. + histol. changes	No-effect level	Remarks
		Inhibition of			Recovery after					
		AChE [d]	ChE [e]	Brain	AChE	ChE	Brain			
Rat 250/250	1/2/3/5/7.5 p.p.m. daily, 7 days a week	3/5/7.5 p.p.m.	7.5 p.p.m.	3/5/7.5 p.p.m.	28 days	7 days	28 days	None	2 p.p.m. [c]	No clinical symptoms at all dosages. Food consumption, bodyweight gain, organ weights and ratios, and blood and urine values normal.
Rat 90/90	0.1/0.5/1.25/2.5/5/10 mg./kg./day, 7 days a week	—	—	—	—	—	—	None	2.5 mg./kg./day	No clinical symptoms at all dosages. Blood and urine values normal, adverse dose-correlated effects on food consumption and bodyweight gain in the 5 and 10 mg./kg./day groups.
Dog 24/24	0.05/0.1/0.5/1/2/3/4/5 mg./kg./day, 6 days a week	1–5 mg./kg./day	1–5 mg./kg./day	—	24 days	10 days	—	None	0.5 mg./kg./day	AChE and ChE activities returned to normal during the test period at the 1 mg./kg./day level.
Dog 9/9	0.1/2.5/5.0 mg./kg./day, 7 days a week	—	—	—	—	—	—	None	2.5 mg./kg./day	OP-poisoning symptoms (one female) at the 5 mg./kg./day dosage. No adverse effects on food consumption and bodyweight gain, or blood and urine values.

[a] Industrial Bio-Test Laboratories (1960 a and b; 1961 a, c, e, and f).

[b] Electrometric method of Michel (1949) as modified by Williams et al. (1957).

[c] A daily dietary level of 2 p.p.m. of phosphamidon is equal to 0.2 mg./kg. bodyweight/day for a rat weighing 100 g. (Association of Food and Drug Officials, USA 1959).

[d] Erythrocyte acetylcholinesterase.   [e] Plasma cholinesterase.

Table VIII. *Results of 90- and 120-day toxicity studies with desethylphosphamidon on rats and dogs* [a]

Animals male/female	Dosage levels	Behaviour of cholinesterases [b]						Pathol.+ histol. changes	No-effect level	Remarks
		Inhibition of			Recovery after					
		AChE [d]	ChE [e]	Brain	AChE	ChE	Brain			
Rat 200/200	0.1/0.3/1/3 p.p.m. daily, 7 days a week	1 p.p.m.	>3 p.p.m.	1 p.p.m.	28 days	—	28 days	—	0.3 p.p.m. [c]	Slight inhibition of AChE and brain cholinesterase.
Rat 175/175	0.3/0.5/0.8/1.25/2 p.p.m., 7 days a week	>2 p.p.m.	>2 p.p.m.	>2 p.p.m.	—	—	—	—	2 p.p.m. [c]	—
Rat 30/30	1/5/10/mg./kg. day, 5 days a week	—	—	—	—	—	—	None	1 mg./kg./day	Dose correlated mortalities and OP-poisoning symptoms at 5 and 10 mg./kg./day. Adverse effect on growth at 10 mg./kg./day. Food consumption, blood, and urine values normal.
Dog 12/12	0.025/0.05/0.1/0.2 mg./kg./day, 6 days a week	0.2 mg./kg./day	>0.2 mg./kg./day	—	50 days	—	—	—	0.1 mg./kg./day	Inhibition of AChE in females only.
Dog 9/9	0.2/1/5 mg./kg./day, 7 days a week	—	—	—	—	—	—	None	1 mg./kg./day	Mortality, OP-poisoning symptoms and bodyweight loss at the 5 mg./kg./day dosage level. Food consumption, blood, and urine values normal.

[a] *Industrial Bio-Test Laboratories* (1964 a, b, and c; 1965 d and e).

[b] Electrometric method of Michel (1949) as modified by Williams *et al.* (1957).

[c] A daily dietary level of 0.3 p.p.m. desethylphosphamidon is equal to 0.03 mg./kg. bodyweight/day and that of 2 p.p.m. is equal to 0.2 mg./kg. bodyweight/day for a rat weighing 100 g. (*Association of Food and Drug Officials*, USA 1959).

[d] Erythrocyte acetylcholinesterase. [e] Plasma cholinesterase.

300 g.), was two (1.6 to 2.4) mg./l. of air *(Industrial Bio-Test Laboratories* 1960 c). Male albino rats exposed for 90 minutes to five percent aerosol concentrations of phosphamidon developed slight dyspnoea after ten to 15 minutes, but recovered after 25 to 30 minutes. Others, exposed to a ten-percent aerosol, developed transient dyspnoea and fascicular twitching occurred after 55 to 63 minutes. They recovered two to four hours after the end of exposure (JAQUES and BEIN 1960).

b) *Subacute toxicity*

1. Oral toxicity. — Male and female albino rats (Sprague-Dawley and Charles River) and pure bred Beagle dogs were chosen for subacute oral toxicity studies (90 and 120 days) on phosphamidon and desethylphospha-midon, the aim being first to establish a dosage level which does not affect blood or brain cholinesterase activity in these animals, and second to establish the level which does not produce clinical or other measurable symptoms. The substances were administered to rats together with the daily food or by intubation, and to dogs in gelatine capsules five to seven times a week. The results of these experiments are summarized in Tables VII and VIII.

2. Dermal toxicity. — A subacute percutaneous toxicity study was carried out with phosphamidon in rabbits. The test material consisted of 85 percent technical grade phosphamidon and 15 percent miscellaneous phosphorylated and non-phosphorylated compounds. The test animals were adult male and female albino rabbits of the New-Zealand strain. The test material was applied at two dose levels to abraded and intact skin, using a patch of gauze covered by plastic sheeting held in place by adhesive tape. The entire trunk of the animal was then covered by a sleeve of gauze elastic stockinette. The animals were exposed for six hours/day (five days a week) for three weeks *(Industrial Bio-Test Laboratories* 1965 c). The results of this study are summarized in Table IX.

Table IX. *Results of 21 day subacute percutaneous toxicity study with phosphamidon*

Group no.	No. of animals		Conc. [a] (%)	Skin condition	Remarks
	Male	Female			
Control	5	5	0	Intact	No mortality, no adverse effects on intact or abraded skin. Bodyweight, behavioral reactions, and hematological, biochemical, histopathological, and urine values all normal.
1	5	5	0.125	Intact	
2	5	5	0.125	Abraded	
3	5	5	0.625	Intact	
4	5	5	0.625	Abraded	

[a] Fifteen ml. of solution applied.

3. Inhalation toxicity. — In a 42-day subacute inhalation study 40 adult male and female Wistar albino rats (one control and three test groups) were exposed for four hours daily, five times a week, to a continuous flow of air containing 0.05 and 0.5 mg./cu. m. of phosphamidon. There was temporary inhibition of erythrocyte cholinesterase, but no mortality, no behavioral reactions, and no hematological, biochemical, or histopathological changes *(Battelle Institute* 1965). In another trial one control group and three treated groups (ten male and female Sprague-Dawley albino rats, ten male and female English strain guinea pigs, and two pure-bred Beagle dogs) were subjected for six hours/day and five days a week over a period of 90 days to a phosphamidon aerosol produced by an Ohio Nebulizer [1]. This device produces droplets of diameter 0.5 to three microns, which penetrate into the alveolar spaces *(Industrial Bio-Test Laboratories* 1964 f, Table X).

Table X. *Results of 90-day subacute inhalation toxicity study with technical grade phosphamidon in rats, guinea pigs, and dogs*

Group no.	Phosphamidon (mg./l. of air)	Remarks
Control	0	No mortality, no adverse effects on bodyweight, behavioral reaction, or hematological, bio-chemical, histopathological, and urine values.
1	0.003	
2	0.016	
3	0.125	

c) *Chronic toxicity*

Chronic toxicity was studied in rats and dogs over two years.

1. Two-year feeding study with rats. — A total of 300 weanling rats (150 males, 150 females, Carworth CFN albino strain) was divided into five equal groups, of which one was a control. The experimental groups were fed 0.1, 0.5, 1.25, and five mg. of phosphamidon/kg. daily, mixed with the stock ration. The dietary concentrations were changed periodically to correct for increase of bodyweight. Checks for mortality and abnormal behaviour were made daily, and the food consumption was also recorded. Bodyweight gains were recorded weekly for the first six months of the study and monthly thereafter. Blood studies including determinations of hemoglobin concentrations, hematocrit values, erythrocyte counts, and both total and differential leucocyte counts were carried out on ten animals of each of the control and the five mg./kg./day group after six, 12, 18, and 24 months of feeding. Urine analysis (glucose and albumin concentration) tests for microscopic elements and urinary pH determinations were conducted on pooled urine samples obtained at the same time intervals and

---

[1] *Ohio Chemical and Surgical Equipment Company,* Madison, Wisconsin.

from the same rats. Gross pathological studies were performed with all rats which died during the test period and with those which were sacrificed after 12 months of testing and at the end of the test period. Organ weights, organ-to-body weight ratios, as well as organ-to-brain weight ratios were recorded. The following organs were included: brain, thyroid glands, heart, liver, kidneys, spleen, adrenal glands, and gonads. Complete microscopical examinations were carried out on the following tissues and organs of the control animals and those from the five mg./kg./day group: heart, aorta, trachea, lungs, liver, pancreas, esophagus, stomach, small intestine, caecum, colon, spleen, lymph nodes, kidneys, urinary bladder, gonads, salivary glands, thyroid glands, parathyroid glands, skeletal muscles, sternum, bone, peripheral nerve, spinal cord, and brain (cerebrum, cerebellum, and pons). All nerve tissues were stained with Luxol Fast Blue, the other tissues with Hematoxylin-Eosin. Tumor incidence, location, weight, size, and pathological classification were recorded for individual animals during and at the end of the above-mentioned investigation. The results were as follows: A significant weight depression (both sexes) occurred only in the five mg./kg./day dose group, and this group consumed less food. No significant abnormalities were observed with respect to survival, behavioral patterns, hematology and urology, organ weights and ratios, gross and microscopical pathological studies, and tumor incidence. There were no signs of poisoning in the second highest dosage group, receiving 1.25 mg./kg./day (*Industrial Bio-Test Laboratories* 1966 a).

2. **Two-year oral study with dogs.** — A control group and three test groups, each consisting of four (two male and two female) pure-bred Beagle dogs were used in the present study. The dosage levels were 0.1, 2.5, and five mg./kg./day. The calculated doses of phosphamidon were prepared weekly in gelatine capsules and were given daily to the dogs, immediately after the normal stock ration. Mortality, clinical symptoms, body weight, and food consumption were checked daily throughout the test period. Complete hematological studies (hemoglobin, hematocrit, erythrocyte, and leucocyte counts), biochemical investigations (blood urea nitrogen, glucose, sulfobromophthalein test), and urine analyses (albumin, glucose, microscopic elements) were carried out with all dogs before the experiment was started, and six, 12, 18, and 24 months after the beginning of the test. At the end of the investigation, all dogs were sacrificed, and the major tissues and organs were examined macroscopically and microscopically. The weights of the following organs were determined: brain, thyroid glands, heart, liver, kidneys, spleen, adrenal glands, and gonads. In the highest dosage group, all dogs died during the test period between the 100th and 600th days, the deaths being attributed to ingestion of the test material which caused symptoms typical of cholinesterase inhibition (e. g., muscular tremors, motor ataxia, salivation, emesis, purulent ocular discharge, and alterations in organ weights and ratios). Moderate clinical symptoms, attributable to cholinesterase inhibition, were also observed at the dose level of 2.5 mg./kg./day. None of the dosages used caused adverse changes in

hematology, clinical blood chemistry, urine, or organ function, and no gross pathological and histopathological changes attributable to phosphamidon were detected. A daily dose of 0.1 mg. phosphamidon/kg./day did not produce changes of any of the parameters investigated (*Industrial Bio-Test Laboratories* 1964 e).

d) *Special toxicity data*

Special toxicological trials, covering skin and eye irritation, neurotoxicity, potentiation, reproduction, cholinesterase inhibition, and the effect of antidotes provided additional toxicological data which support the acute, subacute, and chronic toxicity findings given above.

1. Skin irritation. — The procedure chosen was a modification of a method of DRAIZE *et al.* (1944), using four young adult New Zealand strain albino rabbits with an average body weight of about three kg. A volume of 0.5 ml. of undiluted technical grade phosphamidon was patched on either an intact or abraded skin site on each of the rabbits. The combined average skin irritation score was 1.6 out of a possible maximum of eight, and a rating of "slight irritation" was given (*Industrial Bio-Test Laboratories* 1960 c).

2. Eye irritation. — The procedure chosen was that of DRAIZE *et al.* (1944), using five young adult albino rabbits of the New Zealand strain. A volume of 0.1 ml. of undiluted technical grade phosphamidon was instilled into the conjunctival sac of the right eye of each test animal, and the left served as a control. Undiluted phosphamidon was rated moderately irritating (*Industrial Bio-Test Laboratories* 1960 c).

3. Neurotoxicity. — Certain organophosphates are known to cause a neurotoxic response (prolonged and delayed locomotor ataxia) in animals (BARNES and DENZ 1953, DURHAM *et al.* 1956, DAVIES *et al.* 1960). Neurotoxicity studies with phosphamidon were carried out, therefore, using rats, dogs, and hens as test animals. These experiments are summarized in Tables XI to XIII.

Table XI. *Effect of technical grade phosphamidon on the myelin sheath of various nerve fibers in rats and dogs* [a]

Animal species male/female	Dosage level (mg./kg.)	Substance application	Duration (days)	Remarks
Rat 5/5	20	Feeding study	14	No gross pathological abnormalities. No myelin sheath degeneration of central (cerebrum, cerebellum, pons and lower thoracic area of spinal cord) and peripheral (longitudinal and sciatic nerves and brachial plexus) nervous system.
Dog 1/1	10 (first 4 days) 5 (following 10 days)	Gelatine capsules	14	

[a] *Industrial Bio-Test Laboratories* (1961 g and h).

Table XII. *Results of acute oral toxicity determinations of technical grade phosphamidon and tri-ortho-cresylphosphate (TOCP) to hens (intubation)* [a]

Dose (mg./kg.)	No. of animals tested	No. of animals dead	Observation period (days)	Remarks
10.3	5	0	21	No clinical symptoms.
15.4	5	0	21	Slight to severe symptoms of cholinesterase inhibition. Survivors recovered completely. No myelin sheath degeneration.
23.1	5	3	21	
30.4	5	5	0	
45.6	5	5	0	
1,025	5	5	0	
150 [b]	6	0	21	Symptoms of TOCP poisoning: walking on ankles, complete paralysis of the legs, wry neck. Myelin sheath degeneration.
300 [b]	7	5	21	

[a] *Industrial Bio-Test Laboratories* (1962).
[b] TOCP given for five successive days.

Table XIII. *Results of subacute feeding studies of technical grade phosphamidon and tri-ortho-cresylphosphate (TOCP) to hens* [a]

Dose (p.p.m.)	No. of animals tested	No. of animals dead	Remarks
3	10	1	No clinical symptoms.
60	10	1	
300	10	1	Sporadic severe symptoms of cholinesterase inhibition. No myelin sheath degeneration.
750	10	3	
500 [b]	10	3	At higher dosage levels typical symptoms of TOCP poisoning: Walking on ankles, leg paralysis, wry neck, varying degrees of prostration, typical myelin sheath degeneration.
1,000 [b]	10	2	
3,000 [b]	4	1	
4,000 [b]	4	1	

[a] *Industrial Bio-Test Laboratories* (1962).
[b] TOCP.

4. Potentiation. — Potentiation studies were carried out with a total of 2,018 male Sprague-Dawley albino rats to determine a possible potentiating effect of phosphamidon and desethylphosphamidon on other insecticides (Table XIV). The acute oral toxicity of phosphamidon and desethylphosphamidon was not appreciably potentiated by any of 20 organic phosphates and carbaryl *(Industrial Bio-Test Laboratories* 1961 b, 1965 f, 1966 c, f, and d).

Table XIV. *Comparison of theoretical and observed acute oral LD$_{05}$ values obtained with phosphamidon versus 20 other organic insecticides on rats*

Organic insecticides	LD$_{50}$ (mg./kg.)		
	Theoretical	Observed	Ratio of theoretical/ observed
Phosphamidon/dioxathion	53.7	65	0.83
Phosphamidon/diazinon	80.2	80	1.00
Phosphamidon/naled	214.2	230	0.93
Phosphamidon/ethion	63.2	68	0.93
Phosphamidon/EPN	40.6	44.2	0.92
Phosphamidon/azinphos-methyl	23.6	28.2	0.84
Phosphamidon/malathion	689.2	680	1.01
Phosphamidon/parathion-methyl	23.6	25.2	0.94
Phosphamidon/parathion	18.3	19.9	0.92
Phosphamidon/mevinphos	19.4	14.6	1.33
Phosphamidon/carbaryl	161.7	177	0.91
Phosphamidon/schradan	21.2	29	0.73
Phosphamidon/demeton	18.4	20	0.92
Phosphamidon/carbophenothion	86.7	94	0.92
Phosphamidon/disulfoton	10.2	16.1	0.63
Phosphamidon/coumaphos	58	88	0.66
Phosphamidon/trichlorfon	380	270	1.41
Phosphamidon/fenchlorphos	559	690	0.81
Phosphamidon/DEF® *a*	177	288	0.61
Phosphamidon/dimethoate	185	185	1.00

a S,S,S-Tributyl phosphorotrithioate.

5. Reproduction. — Three successive generations of Wistar-derived albino rats received a phosphamidon mixture (75 percent phosphamidon, 22 percent desethylphosphamidon, and three percent N,N-diethyl-α-chloroacetoacetamide) in accordance with a procedure described by the *Association of Food and Drug Officials*, U.S.A. (1959). There were two control and three test groups (receiving one, 7.5, and 15 p.p.m. of the substance in the daily food). Fifteen p.p.m. in the daily food equals 1.5 mg./kg. for a rat weighing 100 g. *(Association of Food and Drug Officials*, U.S.A. 1959). In the parental animals of all three generations behaviour, mortality, body weight, and rate of weight gain were unaffected, and there were no pathological changes in the tissues which could be attributed to the test material. At all phases, the reproductive performances (mating indices, fertility indices, incidence of pregnancy, and parturition and gestation times) of control and test animals were essentially comparable. Lactating indices ranged from 70 to 100 percent in control groups and from 64 to 100 percent in the 15 p.p.m. dosage group. However, in the second litter of the first generation and both litters of the third generation lactation indices for the animals of the 15 p.p.m. dosage group were slightly lower than the corresponding control values. These values for the one and 7.5 p.p.m. test groups were comparable with those of control animals. There were no outstanding differences between the test and control animals in any of the three generations in respect to the number of young born, the number of stillborn young, and the number of viable young at selected points in the lactation period (up to weaning at 21 days). The body weight of weanlings in the test groups was in all cases comparable with that of weanlings in the control groups. Pathology examination of ten male and ten female weanlings (randomly selected) of all control and test groups of the F3b litters did not reveal tissue alterations which could be correlated with test material ingestion *(Industrial Bio-Test Laboratories* 1965 a and g, 1966 b).

6. **Inhibition of cholinesterases.** — In vitro I_{50} determinations with phosphamidon, desethylphosphamidon, and γ-chlorophosphamidon on different types of cholinesterases (human plasma, horse plasma, bovine erythrocytes, rat brain, and fly head) have been reported by several authors (Bull *et al.* 1967, Jaques and Bein 1960), but the results are not directly comparable because of the differences between the methods. Therefore, I_{50} determinations with pure phosphamidon, desethylphosphamidon, and γ-chlorophosphamidon were repeated under identical experimental conditions with human plasma and bovine erythrocytes as enzyme sources. The findings (Table XV) show that γ-chlorophosphamidon is a more potent cholinesterase inhibitor *in vitro* than pure phosphamidon, but this high anticholinesterase activity is not paralleled by a high acute oral toxicity in mice and rats (see Table III). The discrepancy is due to a very high rate of breakdown of γ-chlorophosphamidon by liver enzymes *(CIBA* 1968).

Table XV. *Inhibition of two types of cholinesterases in vitro by phosphamidon, desethylphosphamidon, and γ-chlorophosphamidon* [a]

Compound	Human plasma	Bovine erythrocytes
Phosphamidon pure	$2.2 \times 10^{-6}\ M$	$6.6 \times 10^{-5}\ M$
Desethylphosphamidon	$2.7 \times 10^{-6}\ M$	$4.8 \times 10^{-6}\ M$
γ-Chlorophosphamidon	$1.6 \times 10^{-8}\ M$	$2.8 \times 10^{-6}\ M$

[a] Using the automated technique of Voss and Geissbuehler (1967).

In vivo cholinesterase inhibition studies were recently carried out with rats and dogs at Tierfarm AG., Sisseln, Switzerland, phosphamidon being applied by gavage or in capsules. The results of these experiments are summarized in Tables XVI and XVII. The highest dosages caused considerable inhibition of both types of cholinesterase. The enzymes, however, recovered completely within a period of six days. The median dose of 1.2 mg./kg. caused an initial decrease of cholinesterase activities with a definitely shorter recovery period. The lowest dose had no effect, except on male rat plasma cholinesterase 0.5 hours after application *(Biologisch-Medizinische Forschungsstation* 1969 e and f).

Table XVI. *Percentage of activity of blood cholinesterases in rats after oral dosing with three concentrations of technical grade phosphamidon P (pre-treatment activity taken as 100 percent)* [a]

Hours after dosing	Males [b]						Females [b]					
	AChE (mg.P/kg.)			ChE (mg.P/kg.)			AChE (mg.P/kg.)			ChE (mg.P/kg.)		
	0.12	1.2	12.0	0.12	1.2	12.0	0.12	1.2	12.0	0.12	1.2	12.0
0.5	112%	78%	26%	54%	12%	29%	116%	73%	30%	82%	35%	28%
3	108%	90%	28%	83%	68%	32%	107%	72%	22%	98%	75%	32%
48	94%	89%	76%	86%	65%	68%	106%	83%	72%	85%	81%	43%
72	106%	95%	—	100%	106%	—	96%	85%	—	100%	99%	—
144 [c]	95%	78%	87%	72%	91%	104%	88%	86%	88%	105%	113%	100%

[a] Method according to Voss and SACHSSE (1970).
[b] Fifteen animals.
[c] Cholinesterase determinations continued over a total period of nine days, values normal.

7. Antidote effect of atropine and PAM.

— Mice injected subcutaneously with phosphamidon (lethal dose of ten mg./kg.) were protected by intraperitoneal administration of atropine (three to 30 mg./kg.) five minutes later. Pyridine-2-aldoxime methiodide (PAM), injected at a dose level of 30 to 70 mg./kg. was also protective, but the best results were achieved by a combined treatment with both substances (JAQUES and BEIN 1960). These results are in agreement with those obtained by ERDMANN *et al.* (1958) with various other cholinesterase inhibitors.

Table XVII. *Percentage of activity of blood cholinesterases in Beagle dogs after oral dosing with three concentrations of technical grade phosphamidon P (pre-treatment activity taken as 100 percent)* [a]

Hours after dosing	Males [b]						Females [b]					
	AChE (mg.P/kg.)			ChE (mg.P/kg.)			AChE (mg.P/kg.)			ChE (mg.P/kg.)		
	0.12	1.2	18.0	0.12	1.2	18.0	0.12	1.2	18.0	0.12	1.2	18.0
0.5	98%	76%	32%	98%	87%	22%	108%	54%	21%	89%	88%	22%
3	99%	103%	55%	101%	99%	30%	124%	103%	60%	98%	94%	29%
24	108%	102%	54%	102%	102%	102%	121%	99%	73%	95%	96%	82%
48	102%	104%	68%	105%	103%	111%	130%	94%	79%	93%	96%	89%
72	103%	98%	77%	101%	104%	107%	124%	99%	83%	94%	92%	84%
96 [c]	102%	102%	88%	100%	102%	111%	124%	112%	92%	97%	88%	89%

[a] Method according to Voss and SACHSSE (1970).
[b] Three animals.
[c] Cholinesterase determinations continued over a total period of 12 days, values normal.

III. Toxicity to wildlife and environmental contamination

Toxicity studies were carried out with phosphamidon on livestock, wild mammals, birds, and fish to complete the toxicological picture obtained from the studies in laboratory animals. A knowledge of the overall effect of insecticides on the environment is of increasing importance.

a) *Livestock and wild mammals*

In a first experiment, two cows with body weights of approximately 700 kg. were fed fresh grass three, four, five, and six days after it had been treated with phosphamidon at a rate of 0.8 kg./ha. Manometric determinations of erythrocyte cholinesterase activity during the test period gave no indication of enzyme inhibition (Heuser 1960).

A second experiment was carried out by *Klipfontein Organic Products Corporation Limited* (1966). Four cows were weekly sprayed to run-off with 0.15 and 0.3 percent phosphamidon over a period of six months. They exhibited neither clinical symptoms nor inhibition of erythrocyte cholinesterase.

During large scale aeral spraying trials for the control of gypsy moths in the United States (repeated application of 0.5 and one kg. phosphamidon/ha.), systematic checks were made at intervals for toxic effects on various species of wild mammals, particularly rabbits, mice, and chipmunks. These observations revealed no hazard to wildlife *(California Chemical Company* 1960).

b) *Birds*

The acute oral LD_{50} of a mixture of 80 percent phosphamidon with 20 percent isopropanol plus dye was determined in various birds (Table XVIII). The test material was diluted in water and given by a stomach

Table XVIII. *Acute oral toxicities of phosphamidon on several bird species*

Animal species and sex	No. animals tested	Age (months)	Weight (kg.)	Acute oral LD_{50} (mg./kg.)
Mallard duck, female	12	3	1.105–1.433	3.05 (2.33–4.00)
Chukar partridge, male and female	—	3–5	0.346–0.542	9.07 (8.3–11.3)
Pigeon, male and female	3	—	—	2.0–3.0
Mourning dove, male and female	3	—	—	2.0–4.0
Whitewing dove, male and female	10	—	0.119–0.167	2.34

tube or by micropipetting onto corn starch in gelatine capsules after starving the birds overnight. Survivors were held for a 14-day observation period. The symptoms of poisoning, which were much the same in all birds, included hyperexcitability, asynergy, lachrymation, foamy salivation, tachypnoea, the holding of the feathers tight against body, and tetany, with knotted claws. Death was caused by respiratory arrest. Gross autopsies showed no pathological changes (Denver Wildlife Research Center 1966).

Technical grade phosphamidon was tested for its acute oral toxicity on nestlings of the chickadee, Parus ater, and adults of the common sparrow, Passer domesticus. In this study phosphamidon was approximately five times as toxic as for rats (see Table I). Surviving individuals recovered rapidly and completely after the crisis (DITTRICH 1966). In acute dermal toxicity studies in captive male whitewing doves, an 80 percent phosphamidon formulation applied to the feathers only at a dilution of one-in-400 gave an LD_{50} value of 110 mg./kg. body weight (Denver Wildlife Research Center 1965).

Subacute and chronic feeding studies with phosphamidon were carried out with mallard ducks, ring-necked pheasants, and bobwhite quails. They were designed to determine the quantities of test compound producing subacute poisoning and death of at least 50 percent of test animals within ten days, and the quantities producing chronic poisoning and 50 percent mortality within 100 days of the test period. The results are given in Table XIX.

A 2,500 ha. block of forest land in Montana, U.S.A., was sprayed with phosphamidon by aircraft at the rate of one pound/acre (about one kg./ha.) against spruce budworm (Christoneura fumiferana). Observation showed that bird population and activity in the sprayed area dropped to about 1/4 of the pre-spray level, while it increased in an unsprayed area. Some birds, including blue grouse (Dendragapus obscurus), were killed by the insecti-

Table XIX. Toxicity of phosphamidon to mallards, pheasants, and bobwhites [a]

| Birds | Test period (days) | Quantities of phosphamidon causing 50% mortality | | | |
| | | Young | | Adult | |
		p.p.m. in diet	mg./kg. eaten	p.p.m. in diet	mg./kg. eaten
Mallard	10	500	430	200	120
	100	100	2,125	100	220
Pheasant	10	500	315	250	90
	100	—	—	200	70
Bobwhite	10	5	4	50	6
	100	1	3	10	80

[a] DEWITT et al. (1963).

cide. Of two sick blue grouse caught and kept in captivity, one died and the other recovered. There was a marked inhibition of cholinesterase activity in both these birds, but it returned to normal in the one which survived (FINLEY 1965). During the course of a large-scale aerial spraying operation with technical grade phosphamidon (approx. 0.25 kg./ha./two l. of water) against the swaine jack-pine sawfly *(Neodiprion swainei Midd.),* in Quebec, Canada, observations were made on the effect of the spray on bird populations in jack pine forests. In spite of the apparent differences in the pre-spray bird populations, it is evident that the application of the spray resulted in a notable reduction in the number of warblers (P < 0.05); other species were, as a group, not significantly reduced in numbers. At half the above rate of application, phosphamidon was still equally effective against the sawfly, and application of this lower rate could reduce the harm to birds (MCLEOD 1967). Phosphamidon was also applied by ground equipment in 16 acres of mature lemon grove in Arizona, as a 0.25 percent aqueous solution in citrus orchards, while a similar four-acre grove nearby was left as an untreated control. These groves are the favoured nesting areas of morning doves; there were 143 nests in the treated area, compared with 51 in the control area. There were no symptoms of acute poisoning in adult or nestling morning doves detected at any time after the application of phosphamidon, and no dead nestlings were found in either area. Other wildlife was also apparently unaffected *(Denver Wildlife Research Center* 1965).

c) *Fish*

Technical grade phosphamidon was used to determine the median tolerance limit (TLm — a concentration which kills 50 percent of the fish) in fathead minnows *(Pimephales promelas,* Rafinesque). Test concentrations from 100 to 1,000 p.p.m. were used, and the temperature, dissolved oxygen, and pH were regularly monitored. The results of the test are given in Table XX.

The acute 24 hour LC_{50} of technical grade phosphamidon at 13° C. to rainbow trout of length 43 to 58 mm. is reported to be 4.4 mg./l. *(U.S. Department of Agriculture* 1966).

Table XX. *Median tolerance limit (TLm) of technical grade phosphamidon on fathead minnows* [a]

Time (hours)	TLm conc. (p.p.m. v/v) [b]
4	> 400
24	400
48	275
96	220

[a] WELLS (1969). [b] Concentrations from 100 to 1,000 p.p.m. v/v were selected.

During an aerial spray application program to control the gypsy moth in Maine, U.S.A., technical grade phosphamidon was used twice at a rate of one kg./ha. The area was covered by a mixed stand of deciduous and coniferous trees. Two pails containing five suckers and five killifish each were placed in an area before spraying commenced, and one pail of similar content was placed in an untreated area. In addition three pails, each containing ten goldfish, were placed in the same way. The fish in the sprayed area were unaffected *(California Chemical Company* 1960).

In trials conducted by the Plant Protection Department of the East Pakistan Government, all fish *(Heteropneustes fossilis* Mueller) survived in water contaminated with 100 p.p.m. of phosphamidon *(CIBA* 1964 b).

Large-scale forest spraying with phosphamidon against spruce budworm in New Brunswick, Canada, at the rate of 0.5 to one kg./ha. did not harm such insects as *Ephemeroptera, Chironomidae,* and *Trichoptera,* which are important as fish food. The numbers of these insects emerging daily in cages set up in two streams, or which drifted onto vertical screens set up in the streams, indicated no changes attributable to phosphamidon (GRANT 1967).

IV. Discussion

The toxicological action of phosphamidon, like that of all other organic phosphates, results from disturbance of nerve function, due to prevention of the hydrolysis of acetylcholine by the enzyme acetyl cholinesterase. The typical clinical symptoms are those of over-excitation of the cholinergic system (DURHAM *et al.* 1962).

On the basis of the acute oral LD_{50} of approximately 20 mg./kg. to the rat, phosphamidon, and desethylphosphamidon are classified as highly toxic compounds, under HAYES' "proposed categories of hazard based on toxicity of economic poisons" (HAYES 1969). In the rat, phosphamidon has an acute LD_{50} five times higher than that of parathion and mevinphos, or the organochlorine insecticide endrin (EDSON *et al.* 1965). The values for dichlorvos and dimethoate are, on the contrary, four and 12 times more favourable (KLIMMER 1963).

The dermal and inhalation toxicities are equally important for the toxicological evaluation of an "economic poison", particularly with regard to the handling of the product under practical conditions. Most acute dermal LD_{50} values for phosphamidon in rats and rabbits lie between 250 and 550 mg./kg., depending on the method of application. Compared with its acute oral toxicity, the dermal toxicity of phosphamidon is relatively favorable. It seems that the rate of resorption of phosphamidon by the skin of warm-blooded animals is relatively low because of the limited liposolubility of the compound. The acute inhalation toxicity (LC_{50}) to the rat of two mg./l. of air after a treatment of four hours can also be considered as favourable. By way of comparison, oxydemetonmethyl has an acute dermal LD_{50} of 162 to 250 mg./kg. for the rat, and an acute inhalation toxicity (LC_{50}) of 1.5 mg./l. of air after a treatment of one hour (Klimmer 1963). Within Hayes' "proposed categories" mentioned above, phosphamidon falls in the second category (toxic) with respect to dermal and inhalation toxicities.

Under practical conditions, the toxicity of the formulations Dimecron 20, 50, and 100 is closely related to that of the technical grade material; i. e., the adjuvants used in the formulation do not much affect the resorption by the oral, dermal, and respiratory routes.

Judging from poisoning symptoms and the death rate, rats and dogs are more sensitive to desethylphosphamidon (subacute oral application) than to phosphamidon. Rats which received five and ten mg. of phosphamidon/kg. over 90 days consumed slightly less food, and gained weight more slowly than controls or test rats receiving the lower dosage levels. Rats and dogs which were treated with the same dosages of desethylphosphamidon developed severe clinical symptoms and died. On the basis of these tests a "no effect level" of 0.1 mg./kg. can be allotted to phosphamidon and desethylphosphamidon in regard to inhibition of blood and brain cholinesterase. In respect to clinical symptoms, however, a "no effect level" of 2.5 mg./kg. can be given for phosphamidon and one mg./kg. for desethylphosphamidon.

Two-year oral studies with phosphamidon on rats and dogs showed a marked difference of sensitivity over this period. Cholinesterase activity was not determined. Rats tolerated a daily dosage of 1.25 mg./kg. of phosphamidon without developing detectable symptoms. Rats receiving five mg./kg. consumed less food and gained weight more slowly compared with those of the control and the other dosage groups. The application of phosphamidon for two years had no effect on tumors which had developed in rats during this period. No differences in respect to the number, site, weight, and nature of the tumors were observed between the rats of the control and those of the treated groups. In comparison, dogs tolerated only a daily dosage of 0.1 mg. of phosphamidon/kg. over the same period without showing symptoms of poisoning. A daily dosage of 2.5 mg./kg. was sufficient to produce moderate clinical symptoms of poisoning. After a daily dosage at five mg./kg., all dogs developed severe symptoms of cholinesterase inhibition of the central and peripheral nervous systems, and died.

These subacute and chronic studies with phosphamidon in rats and dogs point to a "no effect level" of 0.1 mg./kg., which is higher than, for example, the value for coumaphos, and comparable to that for dioxathion (World Health Organisation 1969).

In a special toxicological feeding study, it was evident that 7.5 p.p.m. of a phosphamidon mixture consisting of phosphamidon, desethylphosphamidon, and α-chloroacetoacetdiethylamide, which is equal to 0.75 mg./kg. for a rat weighing 100 g. (Association of Food and Drug Officials, U.S.A. 1959), did not affect either parental rats or their progeny in a three-generation reproduction study. In particular, there was no teratogenicity.

When phosphamidon was administered to rats in combination with 20 commercial cholinesterase inhibiting insecticides (19 organophosphorus compounds and one carbamate), including malathion, there was no significant potentiation. Pre-treatment of mice with sodium phenobarbital, which acts as a liver microsomal enzyme-inducing agent, protected them from poisoning by phosphamidon, dicrotophos, and their N-dealkylated analogs, but dramatically increased the toxicity of dimethoate. On the other hand, the toxicity of malathion to mice was essentially unaffected by pre-treatment with sodium phenobarbital (MENZER and BEST 1968). Furthermore, it is reported that phosphamidon is not a neurotoxic agent which produces myelin sheath degeneration in animals and man, like TOCP or fluorine-containing organophosphorus compounds (DAVIES et al. 1960).

On the basis of all toxicological results in mammals the *Food and Agriculture Organisation of the United Nations, World Health Organization* (1969), estimated an "acceptable daily intake" for man of zero to 0.001 mg. phosphamidon/kg. bodyweight.

Phosphamidon was found to be highly toxic for birds both in the laboratory and in the field. When phosphamidon was fed over a period of ten or 100 days to mallards and pheasants, it was about as toxic for them as lindane, malathion, thimet, and toxaphene. For bobwhites, phosphamidon was much more toxic than the other compounds (DE WITT et al. 1963). In acute oral studies, phosphamidon was found to be one of the most toxic compounds of 30 pesticides studied, especially for mallards, in which it was as toxic as parathion and strychnine *(Denver Wildlife Research Center* 1966). DITTRICH (1966) reported that acute oral application of phosphamidon was highly toxic for chickadees and sparrows, and dimethoate likewise. With regard to toxicity, lindane was between these compounds and DDT. It must be remembered that in field trials the amount of a pesticide to which different birds are exposed varies tremendously, depending on their habits. They may pick up insecticides by dermal contact, inhalation, and ingestion, and many species increase the exposure by ingestion of insecticides with the food or water over a period of days or weeks.

For fish and for the aquatic insects which are their main food, phosphamidon is reported to be less toxic. The toxicity of phosphamidon for coho salmon fry is only 1/250 of that of DDT *(Fisheries Research Board of Canada* (1962).

Summary

We have reviewed extensive toxicological studies in laboratory animals and in the field. Phosphamidon *per os* is highly toxic to rats, but less toxic when inhaled or applied to the skin.

Subacute and chronic toxicity studies with phosphamidon in rats and dogs demonstrate a "no effect level" of 0.1 mg./kg. The symptoms encountered in these studies are those characteristic of cholinesterase inhibition. For desethylphosphamidon, the same "no effect level" and mode of action were established after three months studies in rats and dogs. It seems that desethylphosphamidon has clinically a stronger effect on these animals than phosphamidon itself. Phosphamidon was not carcinogenic to rats when given orally over a two-year period. In a three-generation study in rats, dosage levels of up to 0.75 mg./kg. had no adverse effects on the parents or their progency. When phosphamidon was administered to rats with 19 other organophosphorus compounds and carbaryl, there was no potentiation. Pre-treatment with sodium phenobarbital decreases the toxicity of phosphamidon for mice. In special studies, phosphamidon was found not to be neurotoxic to rats, dogs, and hens. A combination of atropine and PAM can be considered a suitable antidote for phosphamidon poisoning.

Environmental contamination and laboratory studies demonstrate that phosphamidon is highly toxic to birds, but less toxic to fish and to the aquatic insects which are their food.

Résumé *

Toxicologie du phosphamidon

Le présent article passe en revue les études toxicologiques effectuées de manière approfondie, sur des animaux de laboratoire et dans les champs.

Le phosphamidon administré par voie orale est hautement toxique pour les rats, mais il l'est moins lors de son ingestion par inhalation et lors de son application sur la peau. Des études sur la toxicité subchronique et chronique du phosphamidon sur les rats et les chiens donnent un taux sans effets toxiques visibles ("no effect level") de 0,1 mg/kg. Les symptômes observés lors de ces études sont caractéristiques d'une inhibition de la cholinestérase. La valeur du taux sans effets visibles, établie pour le deséthylphosphamidon, s'est avérée semblable à celle du phosphamidon après trois mois d'études sur des rats et des chiens. Du point de vue clinique, le deséthylphosphamidon a un effet plus fort sur ces animaux que le phosphamidon lui-même. L'insecticide administré par voie orale pendant une période de deux ans ne s'est par révélé cancérigène. Au cours d'une étude sur trois générations de rats, des doses de 0,75 mg/kg n'ont eu aucun effet contraire, ni sur les parents, ni sur la progéniture.

La toxicité du phosphamidon n'a pas été augmentée lors de son administration à des rats, avec 19 autres produits organophosphorés et le carbaryle. Un traitement préventif au phénorbital diminue la toxicité du phosphamidon pour les souris. Durant des études spéciales, on n'a pas décelé d'effet neurotoxique du phosphamidon chez les rats, les chiens et les poules. Un mélange d'atropine et de PAM peut être considéré comme un antidote valable contre l'empoisonnement au phosphamidon.

Des observations en plein champ et des études de laboratoire ont montré que le phosphamidon est assez toxique pour les poissons, ainsi que pour les insectes aquatiques qui constituent leur nourriture.

---

* Traduit par J. P. Lang.

Zusammenfassung *
Toxikologie des Phosphamidons

Das vorliegende Kapitel über die Toxikologie des Phosphamidons stützt sich auf Untersuchungen an Laboratoriumstieren und im Freiland. Die Substanz besitzt nach oraler Gabe eine hohe Toxizität, ist aber nach dermaler und inhalatorischer Applikation weit weniger giftig. Der "no effect level", durch subakute und chronische Toxizitätsuntersuchungen an Hunden und Ratten ermittelt, beträgt 0,1 mg/kg. Die beobachteten klinischen Symptome waren in diesen Versuchen charakterisatisch für eine Cholinesterase-Hemmung. Für Desäthylphosphamidon wurden in drei Monate dauernden Versuchen an Ratten und Hunden der gleiche "no effect level" und die gleiche Wirkungsweise festgestellt; verglichen mit Phosphamidon scheint allerdings ein stärkerer klinischer Effekt zu bestehen. Eine über einen Zeitraum von zwei Jahren an Ratten durchgeführte orale Applikation von Phosphamidon führte zu keiner vermehrten Tumorhäufigkeit. Reproduktionsversuche an Ratten (drei Generationen) ließen bis zu einer Dosis von 0,75 mg/kg keine nachteiligen Wirkungen für Elterntiere und deren Nachkommen erkennen. Phosphamidon zeigte auch in Kombination mit 19 anderen Phosphorsäureestern und Carbaryl keine potenzierenden Eigenschaften. Nach Vorbehandlung mit Barbituraten reagierten Mäuse auf Phosphamidon weniger empfindlich. Die Substanz zeigt keine neurotoxische Wirkung an Ratten, Hunden und Hühnern. Als Antidot bei einer Phosphamidon-Vergiftung kann eine Kombination von Atropin und PAM empfohlen werden. Freiland- und Laboratoriumsversuche ergaben, daß Phosphamidon eine hohe Vogeltoxizität besitzt, aber wenig giftig ist für Fische und im Wasser lebende Insekten, die als Fischnahrung dienen.

References

Association of Food and Drug Officials, U.S.A: Appraisal of the safety of chemicals in foods, drugs and cosmetics (1959).

BARNES, J. M., and F. A. DENZ: Experimental demyelination with organo-phosphorus compounds. J. Pathol. Bacteriol. **65**, 597 (1953).

Battelle Institute, Frankfurt (Germany): Investigations on the toxicological effects of repeated inhalation of preparation 12849 (phosphamidon) by rats over a period of six weeks. Unpublished (1963).

Biologisch-Medizinische Forschungsstation, Tierfarm A. G., Sisseln, Switzerland: Report on the acute dermal LD_{50} on the rat with phosphamidon technical. Unpublished (1969 a).

— Report on the determination of the acute intraperitoneal LD_{50} on the mouse with phosphamidon technical. Unpublished (1969 b).

— Report on the determination of the acute oral LD_{50} on the rat with Dimecron 20, 50 and 100. Unpublished (1969 c).

— Report on the determination of the acute dermal LD_{50} on the rat with Dimecron 20, 50 and 100. Unpublished (1969 d).

— Report on the determination of the acute oral LD_{50} and the inhibition of cholinesterases in the rat with phosphamidon technical. Unpublished (1969 e).

---

* Übersetzt von den Autoren.

Biologisch-Medizinische Forschungsstation, Tierfarm A. G., Sisseln, Switzerland: Report on the determination of the acute oral LD_{50} and the inhibition of cholinesterases in the dog with phosphamidon technical. Unpublished (1969 f).

Bucher, K.: Acute oral toxicity test of phosphamidon on rats. Report submitted by the Pharmacological Institute, Univ. of Basle, Switzerland. Unpublished (1959).

Bull, D. L., D. A. Lindquist, and R. R. Grabbe: Comparative fate of the geometric isomers of phosphamidon in plants and animals. J. Econ. Entomol. 60, 332 (1967).

California Chemical Co. (now *Chevron Chemical Co.),* Ortho Division, Research and Development Department, Moorestown, New Jersey: Aerial application of phosphamidon to forested areas – Maine. Unpublished (1960).

CIBA Ltd., Basle, Switzerland: Oral toxicity study of desmethylphosphamidon. Unpublished (1964 a).

— Field trials with phosphamidon and other insecticides to control rice stem borer in Pakistan. Unpublished (1964 b).

— Various reports by the toxicology unit. Unpublished (1964–1968).

— Degradation of phosphamidon and related vinyl phosphates by rabbit liver homogenates. Unpublished (1968).

Clemons, G. P., and R. E. Menzer: Oxidative metabolism of phosphamidon in rats and goats. J. Agr. Food Chem. 16, 312 (1968).

Davies, D. R., P. Roland, and M. J. Rumens: The relationship between the chemical structure and neurotoxicity of alkyl organo-phosphorus compounds. Brit. J. Pharmacol. 15, 271 (1960).

Denver Wildlife Research Center, Denver, Colorado, U.S.A.: Toxicity of pesticides to captive wildlife species. Unpublished annual progress report (1966).

— Observations on effects of phosphamidon on nesting mourning doves at Yuma, Arizona. Unpublished annual progress report (1965).

DeWitt, J. B., W. H. Stickel, and P. F. Springer: Wildlife studies, Patuxent Wildlife Research Centre, 1961–1962, pp. 74–96 (1963). In J. L. George (ed.): Pesticide wildlife studies: A review of Fish and Wildlife Service investigation during 1961 and 1962. *U. S. Fish and Wildlife Service,* Circular 167 (1961–1962).

Dittrich, V.: Investigation on the acute oral toxicity of different pesticides on nestlings of Parus ater and adults of Passer domesticus. Z. Angew. Entomol. 57, 430 (1966).

Draie, J. H., G. Woodward, and H. O. Calvery: Methods for the study of irritation and toxicity of substances applied topically to the skin and mucous membrane. J. Pharmacol. Expt. Therap. 82, 377 (1944).

Durham, W. F., T. E. Gaines, and W. J. Hayes: Paralytic and related effect of certain organo-phosphorus compounds. Arch. Ind. Health 13, 326 (1956).

—, and W. J. Hayes: Organic phosphorus poisoning and its therapy. Arch. Environm. Health 5, 27 (1962).

Edson, E. F., D. M. Sanderson, and D. N. Noakes: Acute toxicity data for pesticides (1964). World Rev. Pest. Control 4, 36 (1965).

Erdmann, W. D., F. Sakai, and F. Scheler: Erfahrungen bei der spezifischen Behandlung einer E 605-Vergiftung mit Atropin und dem Esteraseaktivator PAM. Dtsch. med. Wschr. 83, 1359 (1958).

Finley, R. B., Jr.: Adverse effect on birds of phosphamidon applied to a Montana forest. J. Wildlife Management 29, 580 (1965).

Fisheries Research Board of Canada: Toxicity of the insecticide phosphamidon. Annual Report 62, 117 (1961/1962).

Food and Agriculture Organisation of the United Nations, World Health Organisation: 1968 evaluations of some pesticide residues in food. Geneva (1969).

Grant, C. D.: Effects on aquatic insects of forest spraying with phosphamidon in New Brunswick. J. Fisheries Research Board of Canada 24, 823 (1967).

Hayes, H. W.: Proposed rule making. Federal Register, U.S.A. 34 (64), 6106 (1969).

Heusser, H.: Bericht über einen Fütterungsversuch mit Dimecron-bespritztem Gras. Unpublished (1960).

Industrial Bio-Test Laboratories, Inc., Northbook, Illinois: Ninety-day subacute oral toxicity of phosphamidon technical — dogs. Unpublished (1960 a).
— Ninety-day subacute oral toxicity of phosphamidon — rats. Unpublished (1960 b).
— Acute toxicological studies on phosphamidon. Unpublished (1960 c).
— Effects of phosphamidon on cholinesterase activity in the dog. Unpublished (1961 a).
— Potentiation studies — Phosphamidon versus fourteen organic insecticides. Unpublished (1961 b).
— Effects of phosphamidon on cholinesterase activity in the rat. Unpublished (1961 c).
— Ninety-day subacute oral toxicity of phosphamidon — rats. Pathologic findings. Unpublished (1961 d).
— Ninety-day subacute oral toxicity of phosphamidon technical — dogs. Pathologic findings. Unpublished (1961 e).
— Effects of phosphamidon on cholinesterase activity in the dog. Unpublished (1961 f).
— Effects of phosphamidon upon the myelin sheath of various nerve fibers in the albino rat. Unpublished (1961 g).
— Effects of phosphamidon upon the myelin sheath of various nerve fibers in the dog. Unpublished (1961 h).
— Demyelination studies in chickens-phosphamidon. Unpublished (1962).
— Toxicity studies on two materials. Unpublished (1963).
— 14-week subacute oral toxicity of desethylphosphamidon — Beagle dogs. Cholinesterase data. Unpublished (1964 a).
— Fourteen-week subacute oral toxicity of desethylphosphamidon — dogs. Unpublished (1964 b).
— 90-day subacute oral toxicity of desethylphosphamidon — albino rats. Unpublished (1964 c).
— Comparative toxicity studies on phosphamidon and gamma-chlorophosphamidon. Unpublished (1964 d).
— Chronic oral toxicity study of phosphamidon — Beagle dogs. Unpublished (1964 e).
— 90-day subacute aerosol inhalation toxicity of technical phosphamidon. Unpublished (1964 f).
— Progress report — three generation reproduction study on a phosphamidon mixture (75 percent phosphamidon, 22 percent desethylphosphamidon and three percent N,N-diethyl-α-chloroaceto-acetamide). Unpublished (1965 a).
— Oral toxicity study of deschlorophosphamidon. Unpublished (1965 b).
— Repeated dermal toxicity of phosphamidon 8 spray. Unpublished (1965 c).
— Effects of desethylphosphamidon on cholinesterase activity in the Beagle hound. Unpublished (1965 d).
— Effects of desethylphosphamidon on cholinesterase activity in the albino rat. Unpublished (1965 e).
— Potentiation studies — desethylphosphamidon versus fourteen insecticides. Unpublished (1965 f).
— Three generation reproduction study in albino rats — a phosphamidon mixture. Results through weaning of Flb litters. Unpublished (1965 g).
— Chronic oral toxicity of phosphamidon, albino rats. Unpublished (1966 a).
— Final report — three generation reproduction study on a phosphamidon mixture — albino rats. Unpublished (1966 b).
— Phosphamidon potentiation studies — with six organic phosphates. Unpublished (1966 c).
— Desethyl-phosphamidon-potentiation studies with six organic phosphates. Unpublished (1966 d).
— Ninety-day cholinesterase study of SX-199 desethylphosphamidon — albino rats. Unpublished (1969).

JAQUES, R., and H. J. BEIN: Toxikologie und Pharmakologie eines neuen systemisch wirkenden Insektizids der Phosphorsäureester-Reihe, Phosphamidon (2-Chlor-2-diäthyl-carbamoyl-1-methylvinyldimethylphosphat). Arch. Toxikol. **18**, 316 (1960).

KLIMMER, O. R.: Pflanzenschutz und Schädlingsbekämpfungsmittel, Abriß einer Toxikologie und Therapie von Vergiftungen (1963).

Klipfontein Organic Products Corporation Limited, Kempton Park, South Africa: Cholinesterase level determination of animals sprayed with phosphamidon. Unpublished (1966).

Klotzsche, C.: Neue insektizide Phosphor- und Phosphonsäureester. Nachr'bl. deutsch. Pfl'sch'dienstes 10, 60 (1958).

— Zur toxikologischen Prüfung neuer insektizider Phosphonsäureester. Int. Arch. Gewerbepathol. Gewerbehyg. 21, 92 (1964).

Mcleod, J. M.: The effect of phosphamidon on bird populations in jack pine stands in Quebec. Can. Field Naturalist 81, 102 (1967).

Menzer, R. E., and N. H. Best: Effect of phenobarbital of the toxicity of several organophosphorus insecticides. Toxicol. Applied Pharmacol. 13, 37 (1968).

Michel, H. O.: An electrometric method for the determination of red blood cells and plasma cholinesterase activity. J. Lab. Clin. Med. 34, 1564 (1949).

National Hygienic Laboratory, Yoshio, Ikeda: Toxicity test of phosphamidon. Unpublished (1958).

Noakes, D. N., and D. M. Sanderson: A method for determining the dermal toxicity of pesticides. Brit. J. Ind. Med. 26, 59 (1969).

U.S. Department of Agriculture: Suggested guide for the use of insecticides to control insects affecting crops, lifestock, households, stored products, and forest products. Agricultural Handbook No. 313 (1966).

Voss, G., and H. Geissbuehler: Automated residue determination of insecticidal enolphosphates. Med. Rijks faculteit Landbouwwetenschappen, Gent XXXII, 877 (1967).

—, and K. Sachsse: Red cell and plasma cholinesterase activities in microsamples of human and animal blood determined simultaneously by a modified acetylthiocholine/ DTNB procedure. Toxicol. Applied Pharmacol. 16, 764 (1970).

Wells, D. L.: Toxicity evaluation of the insecticide phosphamidon. Ontario Water Resources Commission. Unpublished (1969).

Williams, M. W., J. P. Frawley, H. N. Fuyat, and J. R. Blake: Modification of the Michel electrometric technique for dog and rat blood cholinesterase. J. Assoc. Official Agr. Chemists 40, 1118 (1957).

Chapter 6

The behavior of phosphamidon in plants

By

G. Voss and H. Geissbühler

Contents

I. Introduction

Phosphamidon is a systemic insecticide (BACHMANN 1957, ANLIKER et al. 1961) and as such penetrates into all plant tissues treated, i. e., roots, stems, or leaves. After penetration it is translocated in the plant tissues upward from the point of application. A weak downward flow may favor the distribution of the compound in single leaves, but not in the whole plant.

This brief introduction summarizes the results of experimental work carried out in the past on the behavior of phosphamidon in and on plants. The authors feel that a detailed description of experiments on uptake, penetration, and translocation, performed under controlled laboratory or greenhouse conditions, is essential for a more complete understanding of degradational and metabolic processes as well as of residue data, which are described in separate chapters of the present volume.

II. Translocation after root uptake

The most typical property of a water-soluble systemic pesticide is its relatively fast uptake by the roots, after which the compound is translocated to all other parts of the plants. This type of translocation most

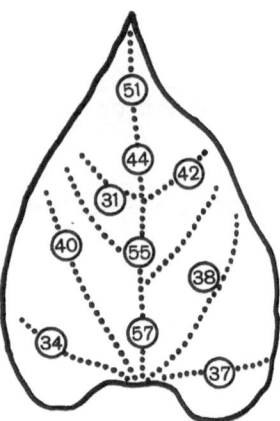

Fig. 1. Primary leaf of bean plant, *Phaseolus vulgaris*, the roots of which had been ex-
    posed for 24 hours to a nutrient solution containing 0.05 percent phosphamidon.
    The ringed numbers indicate percentage of cholinesterase inhibition by phospha-
    midon present in small leaf discs cut from the leaf by means of a cork borer (Voss
    1967, by kind permission of *Duncker und Humblot Verlag,* Berlin)

probably occurs in the xylem, since BENNETT (1949) and TIETZ (1954) observed that the rates of translocation of dimefox and of demeton, respectively, increased with the rate of transpiration. Voss (1967) and Voss and
DITTRICH (1967) demonstrated the presence of phosphamidon in all leaves
of bean, rice, and cotton plants after the roots had been exposed to the
insecticide either in nutrient solution or in soil. Figure 1 shows a primary
leaf of a bean plant, leaf discs of which were analysed by a cholinesterase
inhibition procedure. The concentration of phosphamidon along the main
vein was found to be slightly higher than that in the other parts of the
leaf. Figure 2 a demonstrates that inhibition of cholinesterase occurred with
leaf discs from all leaves of a plant exposed to phosphamidon in a nutrient

Table I. *Concentration of phosphamidon in stems and leaves of bean plants after uptake
by the root system from a nutrient solution containing 500 p.p.m. of insecticide at 25° C.,
and 60 to 70 percent relative humidity (CIBA 1968)*

Hours in nutrient solution with 500 p.p.m. phosphamidon	Phosphamidon (p.p.m.)	
	In stems	In leaves
2	55	15
7	116	101
23	150	200
31	132	398
47	150	476

solution. The amount of the insecticide present in young trifoliate leaves was slightly higher than that in the older, primary leaves. These results were confirmed in parallel bioassay experiments using spider mites *(Tetranychus urticae* Koch) as a test organism (Voss and DITTRICH 1967). In a more detailed experiment, leaves and stems of beans plants, the roots of which had been exposed to a nutrient solution containing phosphamidon, were extracted after various time intervals and analyzed for the insecticide by an automated cholinesterase inhibition method (Voss and GEISSBUEHLER 1967). The results (Table I) showed a steady accumulation of phosphamidon in the leaves, whereas equilibrium was reached in the stems after approximately 20 hours. The insecticide concentration in the leaves equalled that of the nutrient solution after two days. In another experiment, the

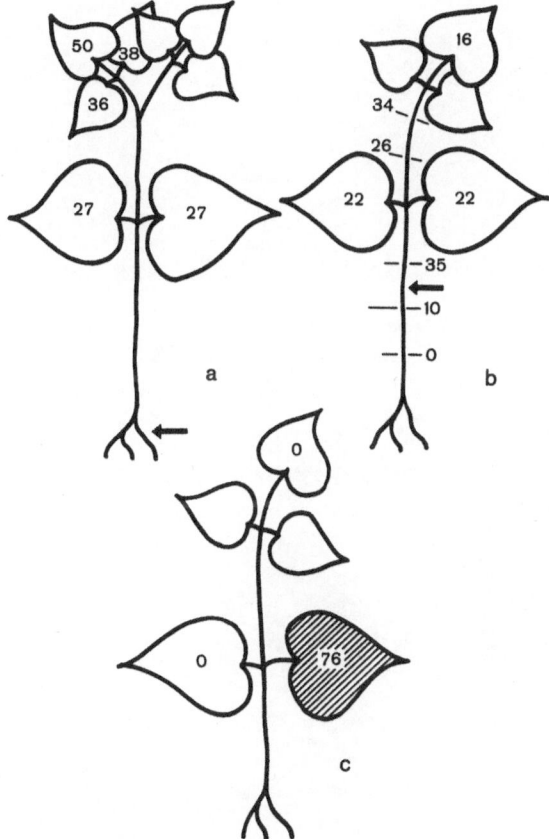

Fig. 2. Schematic drawings of bean plants, treated with phosphamidon in three different ways: *a* = root application, *b* = stem application, and *c* = leaf application. Numbers indicate percentage of cholinesterase inhibition by phosphamidon, present in various parts of the treated plant (leaf discs, stem sections)

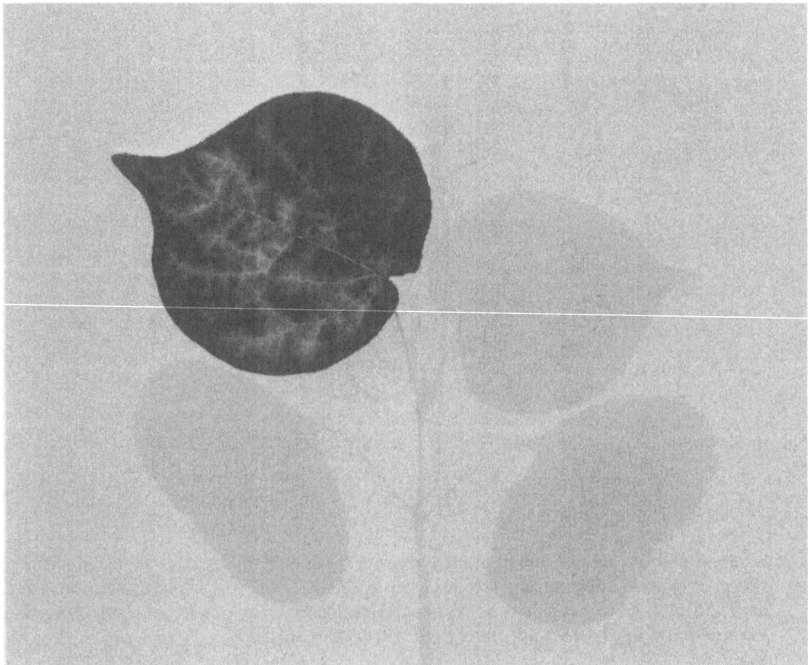

Fig. 3. Autoradiographs of cotton plants, single leaves of which were brushed with a 0.05 percent aqueous solution of ¹⁴C-phosphamidon. Above left and right =

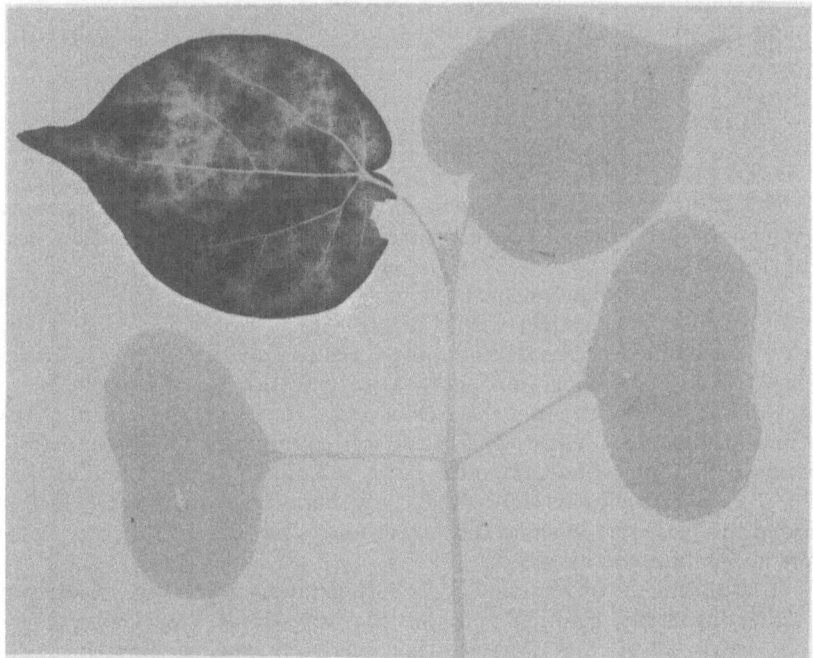

mounted plant specimens, below left and right = autoradiographs, bottom left = three days after application, and bottom right = six days after application

concentration of phosphamidon in bean and rice plants was found to be directly proportional to that in the nutrient solution in which they were grown.

III. Translocation after stem application

Phosphamidon was found to penetrate the stems of bean plants, after which it was rapidly translocated upward (Voss and Dittrich 1967). A typical example is shown in Figure 2 b. Transverse sections of parts of the stem above the point of application, and leaf discs taken from primary and secondary leaves, showed considerable cholinesterase inhibition when assayed 24 hours after application of the insecticide, but almost no inhibition was observed with sections from stem parts below the point of application. Ridgway and Randolph (1968) applied P^{32}-labeled phosphamidon to stems and leaf sheaths of maize plants, and found that treatment of the stems gave better uptake and a more even distribution of radioactivity within the plant.

A further experiment on the translocation of phosphamidon after leaf application was carried out by Bull et al. (1967), who injected P^{32}-labeled phosphamidon into cotton leaves, and later determined radioactivity in various plant parts. Four days after application 87 percent of the applied dose was still present in the treated leaves. The percentage of radioactivity in upper and lower leaves was found to be 2.6 and 0.9, respectively. Terminals contained 1.9 and roots 1.5 percent of total radioactivity, which was shown to be due to breakdown products (dimethyl phosphate and phosphoric acid), but not to the parent compound itself.

IV. Translocation after leaf application

The direction of phosphamidon movement within the plant after leaf application was examined under greenhouse conditions in cotton, bean, and rice seedlings.

In a first set of experiments the entire surfaces of single leaves of young cotton plants were brushed with a solution of C^{14}-phosphamidon. Migration was examined by autoradiography three and six days after application (Fig. 3). The autoradiographs showed no evidence for significant movement of the insecticide from the treated leaf into its petiole, into the stem, or into neighbouring leaves. This result was confirmed in bean plants, using cholinesterase inhibition and mite assay techniques (Voss and Dittrich 1967). Figure 2 c clearly shows that only discs from the treated leaf inhibited cholinesterase whereas discs from untreated leaves of the same plant gave no inhibition at all.

In a second set of experiments, phosphamidon was applied locally by spotting insecticide solution onto the central part of bean leaves (Fig. 4, black area). When discs were punched out from various parts of such a

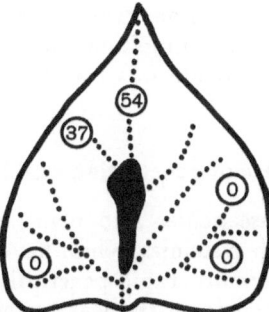

Fig. 4. Bean leaf spotted with 0.1 ml. of a one percent aqueous solution of phosphamidon (black area) and assayed 24 hours later by a cholinesterase inhibition method. Ringed numbers indicate the percentage of cholinesterase inhibition by leaf discs (Voss 1967, by kind permission of *Duncker und Humblot Verlag*, Berlin)

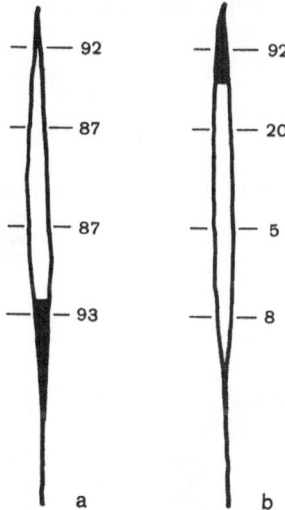

Fig. 5. Leaves of rice plants (25 cm. high) painted with an aqueous solution of phosphamidon (black areas: *a* = base treated, *b* = apex treated). Numbers indicate percentage of cholinesterase inhibition by phosphamidon present in leaf cross sections 24 hours after application (Voss and DITTRICH 1967, by kind permission of *Paul Parey Verlag*, Hamburg)

leaf, the only ones which inhibited cholinesterase were those taken from the apical part of the leaf. The findings were the same when the bases of rice leaves were painted with phosphamidon solution (Fig. 5 a). Small cross sections taken at various distances above the application area showed high cholinesterase inhibition values. However, when rice leaves were painted at the top (Fig. 5 *b*), sections taken at points below the application area exhibited only negligible inhibition, indicating that only a minute fraction of the applied insecticide was translocated downward (Voss and Dittrich 1967).

V. Rate of penetration into leaves

The penetration of insecticides into plant leaves is of practical importance, particularly with systemics, which have to control various plant-sucking arthropods. Furthermore, rapid penetration will help to ensure that an insecticide is not washed from the leaf surface by rain. On the other hand, the fraction of the applied compound which does penetrate into the leaves may disappear rapidly as a result of metabolic breakdown.

The fate of the geometric isomers of P³²-labeled phosphamidon in greenhouse grown cotton leaves after topical treatment was investigated by Bull *et al.* (1967). The ratio of the external to the internal residues of phosphamidon and desethylphosphamidon (both isomers of each) was 2.3

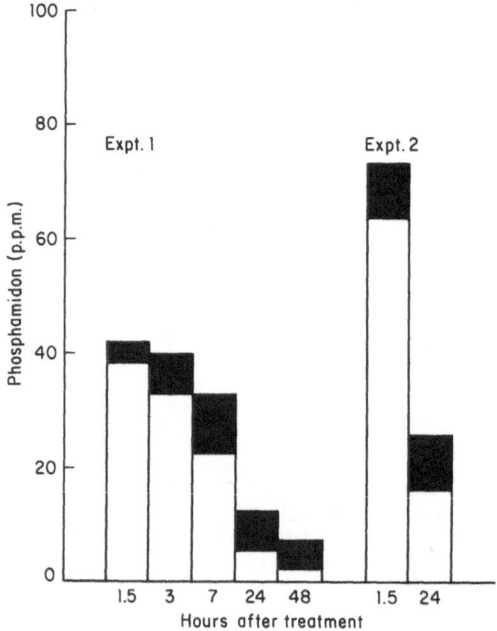

Fig. 6. Rate of penetration of phosphamidon into leaves of *Phaseolus vulgaris*. Experiment 1 = 0.02 percent active ingredient, sprayed; experiment 2 = 0.05 percent active ingredient, dipped. Black parts in columns represent internal residues, white areas external residues. External residues were removed by shaking the intact treated leaves in tap water for three minutes. Residues were analysed by the automated cholinesterase inhibition method

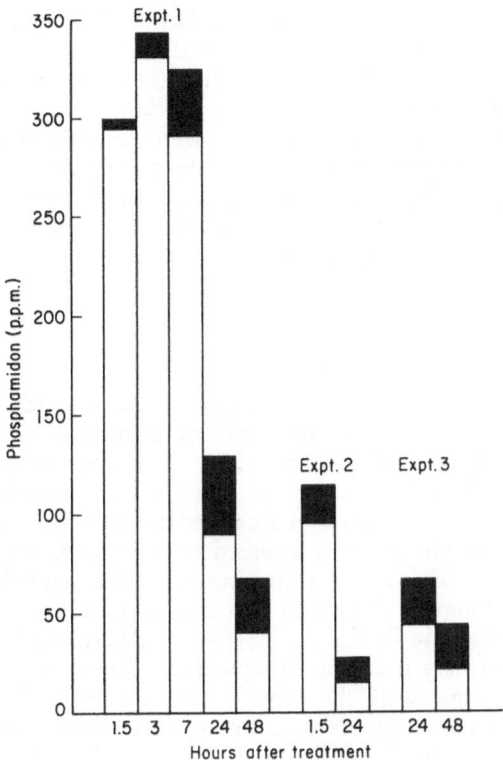

Fig. 7. Rate of penetration of phosphamidon into leaves of rice plants. Experiment 1 = 0.1 percent active ingredient, sprayed; experiment 2 = 0.05 percent active ingredient, dipped; and experiment 3 = ultra-low-volume application. Back parts in columns represent internal residues, white areas external residues. External residues were removed by shaking the intact treated leaves in tap water for three minutes. Residues were analysed by the automated cholinesterase inhibition method

one day after application, and 1.8 two days after application. Comparable results were obtained with an automated cholinesterase inhibition method (Voss and GEISSBUEHLER 1967) for the analysis of external and internal residues of phosphamidon plus desethylphosphamidon (*CIBA* 1968). Figure 6 shows the results of two experiments with potted beans. The penetration of phosphamidon into leaves of rice plants was also studied (Fig. 7). In the latter experiment ultra-low volume application was included, but the rate of penetration did not differ from that observed after high-volume application of aqueous solutions. Penetration into bean leaves appeared to be slightly faster than that into rice leaves. This is shown by the mean external/internal ratios obtained after 24 hours, which were 1.28 and 1.61, respectively. The rate of penetration of phosphamidon into leaves of coffee plants was found to be much lower than that into leaves of beans and rice, the above-mentioned ratio being nine and eight 24 and 48 hours after application, respectively. It is evident from these results that the surface structure and the hardness of leaves play an important role in the penetration of phosphamidon. Similar findings were recently reported by LIND-QUIST and BULL (1967) who investigated the penetration of monocrotophos into cotton leaves. They observed that the rate of penetration was relatively high in young growing leaves, intermediate in mature leaves, and low in old ones with a hard surface.

VI. Discussion

The behaviour of phosphamidon in plants corresponds to that of other well-known systemic insecticides, such as demeton, dimethoate, mono-crotophos, and dicrotophos (TIETZ 1954, DE PIETRI-TONELLI 1966, Voss and DITTRICH 1967). The favorable water/solvent partitioning behavior of phosphamidon, a compound of intermediate polarity, facilitates uptake and translocation. The latter is assumed to occur in the xylem, since first its rate depends on the rate of transpiration, and second the substance accumulates in the marginal parts of leaves as a consequence of the conductive tissue ending blindly in this region. Furthermore, the direction of migration was always found to be upward and outward in various plant species after treatment of the roots, stems, or leaves. The finding that phosphamidon often accumulates in young leaves to a greater extent than in older ones is also explainable on the basis of the higher rate of transpiration observed in growing leaves with a high metabolic activity (BOL-LARD 1960). Thus, a number of phenomena connected with the behavior of systemic insecticides in plants can be explained by comparatively simple physical processes. Since the autoradiographs presented in this publication did not show any evidence of downward movement, we believe that no translocation occurs in the phloem tissue.

The finding that phosphamidon is translocated only upward is of considerable practical importance. If the spray coverage of a plant is inadequate, even distribution of phosphamidon in it cannot be expected. The most obvious advantage a systemic insecticide has to offer is the possibility of successful application to roots and stems. These modes of application have been investigated and used in woody plants with compounds such as demeton or dimefox (WEDDING 1953, HANNA et al. 1955, NORRIS 1965, JOHNSON and REDISKE 1965) and in cotton plants with monocrotophos (RIDGWAY and LINDQUIST 1966).

The rate of penetration of phosphamidon into the leaves of treated plants depends on the surface structure of the particular type of leaf treated. With bean plants a ratio of external to internal residues of one was found after approximately one day. Rice leaves showed a ratio between one and two after the same period, and the penetration into leaves of coffee plants was not at all efficient, the ratio being nine. These results indicate that rainfall shortly after spray applications of the insecticide will normally reduce its efficiency.

Summary

The translocation of phosphamidon in plants was found to occur in an upward direction only. This was not only true with the compound being applied to the roots of plants growing in soil or nutrient solution, but also with phosphamidon brushed or spotted on stems and leaves. The direction of migration of the substance was investigated by using different techniques, such as radio assay, biochemical assay, and biological assay. Phosphamidon also penetrates into leaves after foliar application, but the rate of penetration strongly depends on the leaf age and the structure and properties of the leaf surface. Consequences arising from the translocation and penetration behavior are briefly discussed with regard to practical field applications.

Résumé *

Le comportement du phosphamidon dans les plantes

Le phosphamidon appliqué par les racines ou sur les feuilles et les tiges, subit une translocation exclusivement apicale. Ceci a été mis en évidence par diverses méthodes (phosphamidon radioactif, expériences biochimiques et biologiques). Après un traitement foliaire, le phosphamidon est capable de pénétrer dans les feuilles, mais la vitesse de pénétration dépend fortement de l'âge, de la structure et des propriétés de la surface de celles-ci. Les conclusions qui découlent des propriétés de translocation et de pénétration, seront discutées brièvement en relation avec les différentes applications en plein champ.

Zusammenfassung **

Das Verhalten von Phosphamidon in Pflanzen

Phosphamidon wird in Pflanzen sowohl nach Wurzel- als auch nach Blatt- oder Stengelapplikation nur in Aufwärtsrichtung translociert. Dies wurde mit Hilfe verschiedener Methoden (radioaktiv markiertes Phosphamidon, biochemische und biologische Tests) nachgewiesen. Nach einer Blattbehandlung dringt die Substanz ins Blattinnere ein, wobei die Eindringungsgeschwindigkeit vom Alter und der Oberflächenstruktur des Blattes abhängt. Schlußfolgerungen, die sich aus dem Translokations- und Penetrationsverhalten für die Praxis ergeben, werden kurz diskutiert.

---

* Traduit par les auteurs.
** Übersetzt von den Autoren.

References

Anliker, R., E. Beriger, M. Geiger, and K. Schmid: Über die Synthese von Phosphamidon und seinen Abbau in Pflanzen. Helv. Chim. Acta 44, 1622 (1961).

Bachmann, F.: Phosphamidon, ein neuer Phosphorsäureester mit systemischer Wirkung. Proceed. IVth Internat. Congress Crop Protection (Hamburg) 2, 1153 (1957).

Bennett, S. H.: Preliminary experiments with systemic insecticides. Ann. Applied Biol. 36, 160 (1949).

Bollard, E. G.: Transport in the xylem. Ann. Rev. Plant Physiol. 11, 141 (1960).

Bull, D. L., D. A. Lindquist, and R. R. Grabbe: Comparative fate of the geometric isomers of phosphamidon in plants and animals. J. Econ. Entomol. 60, 332 (1967).

CIBA Ltd., Agrochemical Division, Basle, Switzerland: Weitere Untersuchungen zum Verhalten von Dimecron, Carbicron und Nuvacron in und auf Pflanzen. Unpublished report (1968).

de Pietri-Tonelli, P.: Penetration and translocation of Rogor applied to plants. Adv. Pest Control Research 6, 31 (1966).

Hanna, A. D., E. Judenko, and W. Heatherington: Systemic insecticides for the control of insects transmitting swollen-shoot virus diseases of cacao in the Gold Coast. Bull. Entomol. Research 46, 669 (1955).

Johnson, E. N., and J. H. Rediske: Systemic pesticides in woody plants. Translocation. Bull. Entomol. Soc. Amer. 11, 190 (1965).

Lindquist, D. A., and D. L. Bull: Fate of 3-hydroxy-N-methyl-ciscrotonamide dimethylphosphate in cotton plants. J. Agr. Food Chem. 15, 267 (1967).

Norris, D. M.: Systemic pesticides in woody plants. Uptake. Bull. Entomol. Soc. Amer. 11, 187 (1965).

Ridgway, R. L., and N. M. Randolph: Stem application of systemic insecticides to corn. J. Econ. Entomol. 61, 581 (1968).

Tietz, H.: Der mit ^{32}P markierte Diäthylthionophosphorsäureester des β-oxäthyl-thioäthyläthers (Wirkstoff des systemischen Insektizids „Systox"), seine Aufnahme in die höhere Pflanze und sein Wanderungsvermögen. Höfchen-Briefe, pp. 1–56 (1954).

Voss, G.: Systemische Phosphor-Insektizide. Eine neue Methode zum Studium der Aufnahme und Translokation in Pflanzen. Zeitschr. für angew. Zool. 54, 113 (1967).

—, and V. Dittrich: The translocation of insecticidal enolphosphates in plants. Zeitschr. für angew. Entomol. 59, 430 (1967).

—, and H. Geissbuehler: Automated residue determinations of insecticidal enolphosphates. Mededelingen Rijksfaculteit Landbouwwetenschappen Gent (Belgium) 32, 877 (1967).

Wedding, R. T.: Plant physiological aspects of the use of systemic insecticides. J. Agr. Food Chem. 1, 832 (1953).

Chapter 7

Phosphamidon residue methods

By

G. Voss, I. Baunok, and H. Geissbühler

Contents

I. Introduction

The aim of this review article is to summarize the various methods which have been used at one time or another for qualitative and quantitative determinations of residues of phosphamidon and its major metabolites in plant and animal materials. The methods used over the past ten years were developed in many laboratories all over the world, and cover a wide range of different analytical procedures. Some of these procedures are well known to residue analysts, e. g., enzymatic methods or total phosphorus determinations. Others are recent improvements of known techniques, or new developments, which are described in this article for the first time. This paper will also evaluate the advantages and disadvantages of the methods with regard to sensitivity, specificity, reliability, and ease of operation.

The most important compounds from the point of view of residues are phosphamidon, and its major toxic metabolite N-desethylphosphamidon (see Chapter 4). Residue methods are also available for O-desmethylphosphamidon, N,N-diethyl-chloroacetoacetamide, N-ethyl-chloroacetoacetamide, and gamma-chlorophosphamidon, an impurity present in technical grade phosphamidon.

II. Extraction and methods of cleanup

Among the physico-chemical properties of phosphamidon the complete miscibility with solvents of intermediate to high polarity, such as chloroform, methylene chloride, acetonitrile, and water on the one hand, and its limited solubility in hexane and other aliphatics on the other hand, are particularly helpful in connection with cleanup and separation procedures in residue analysis. These properties allow phosphamidon to be separated from lipophilic plant constituents and/or less polar organophosphates, such as parathion, malathion, diazinon, trithion, bromophos, chlorfenvinphos, and many others, by a simple water-hexane partitioning step, after which phosphamidon is recovered in the aqueous phase. It can then either be used directly for residue analysis, i. e., by cholinesterase inhibition, where the final extract has to be in an aqueous solution, or reextracted from the water by organic solvents of medium polarity, such as chloroform or methylene chloride, and then concentrated for gas chromatographic analysis or thin-layer chromatography (TLC).

Methods for the extraction of phosphamidon, and subsequent separation from interfering plant materials (cleanup) or other organophosphates (specificity) are summarized in Tables I, II, and III. Table II gives the limits of detection and the percentage recoveries achieved with the various methods. Table III summarizes the chromatographic properties of phosphamidon and its metabolites. A detailed description of extraction and cleanup procedures is given in section IV (recommended methods) of the present article.

III. Methods of residue analysis of phosphamidon and metabolites

Methods developed since 1960 for residue analysis of phosphamidon and its metabolites offer a wide range of possibilities, including bioassays (*Daphnia*, cholinesterase), chemical analysis (determinations of diethylamine or total phosphate, and reaction of phosphamidon with nitrobenzylpyridine and blue tetrazolium chloride), and physico-chemical procedures such as infrared spectroscopy and gas chromatography (GLC). These methods are briefly reviewed below.

Table I. *Schematic presentation of extraction, cleanup, and identification procedures used for quantitative and specific determination of phosphamidon*

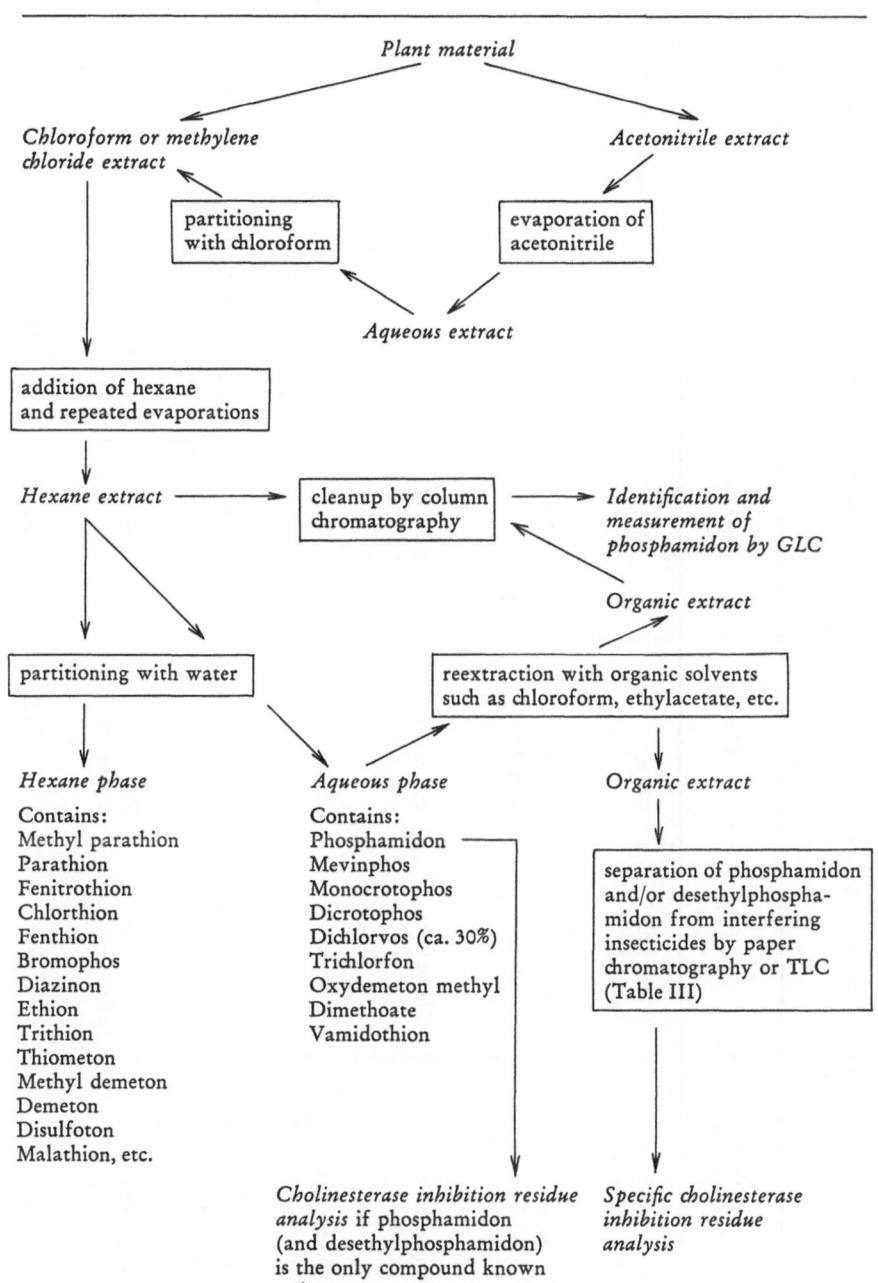

Plant material

Chloroform or methylene chloride extract

partitioning with chloroform

Acetonitrile extract

evaporation of acetonitrile

Aqueous extract

addition of hexane and repeated evaporations

Hexane extract → cleanup by column chromatography → *Identification and measurement of phosphamidon by GLC*

Organic extract

partitioning with water

reextraction with organic solvents such as chloroform, ethylacetate, etc.

Hexane phase

Contains:
Methyl parathion
Parathion
Fenitrothion
Chlorthion
Fenthion
Bromophos
Diazinon
Ethion
Trithion
Thiometon
Methyl demeton
Demeton
Disulfoton
Malathion, etc.

Aqueous phase

Contains:
Phosphamidon
Mevinphos
Monocrotophos
Dicrotophos
Dichlorvos (ca. 30%)
Trichlorfon
Oxydemeton methyl
Dimethoate
Vamidothion

Organic extract

separation of phosphamidon and/or desethylphospha-midon from interfering insecticides by paper chromatography or TLC (Table III)

Cholinesterase inhibition residue analysis if phosphamidon (and desethylphosphamidon) is the only compound known to be present

Specific cholinesterase inhibition residue analysis

Table II. *Summary of methods, including extraction and cleanup procedure, limits of detection, and recovery values, for phosphamidon residue analysis*

Method (final measurement)	Extraction	Cleanup	Limit of detection (p.p.m.)	Recoveries (%)	References
Bioassay with *Daphnia magna*	Water with Tween 80	None	ca. 1.0	Not recorded	Diemair and Knopf (1964)
Cholinesterase (ΔpH method)	Water and hexane	Hexane evaporation in presence of water	0.05–0.1	> 80	Pack et al. (1964)
Cholinesterase (Thiodholine)	Chloroform	Hexane-water partitioning	0.05–0.1	> 80	This paper
Cholinesterase (Automated method)	Chloroform	Hexane-water partitioning	0.05–0.1	> 80	Voss and Geissbuehler (1967)
Diethylamine determination	Not recorded	Not recorded	ca. 1.0	Not recorded	cf. Pack et al. (1964)
Total phosphate	Methylene-chloride	Paper dhromatography	0.5–1.0	> 80	Anliker and Menzer (1963)
Total phosphate	Chloroform	Charcoal column	0.1	> 90	Brewerton (1963)

Method	Solvent	Procedure		Reference
Total phosphate	Acetonitrile	Ethyl acetate-water partitioning, column a, TLC b	0.04 — 97±13	CIBA (1965)
Nitrobenzyl-pyridine	Extraction and cleanup according to STORHERR et al. (1964)		0.1–0.2 — Not recorded	GETZ and WATTS (1964)
Double paper chromatographic method, blue tetrazolium chloride	Methanol	Hexane-water partitioning, reextraction from water with methylene chloride	0.01 — > 80	ANLIKER et al. (1961)
Infrared spectroscopy	Extraction and cleanup according to LAWS and WEBLEY (1961)		0.2 — ~ 70	CROSBY and LAWS (1964)
Gas chromatography	Acetonitrile or methylene chloride	Hexane-water partitioning, silica gel column	0.03 — > 80	This paper

a Activated carbon, magnesium oxide, Celite column.
b To separate phosphamidon from fenitrothion.

Table III. *Summary of chromatographic procedures, detection systems, and R_f-values used for residue analysis of phosphamidon, its major metabolites, and by-product*

Compound	Chromatographic support	Solvent system	Detection system	R_f value	Reference
Phosphamidon	Paper chromat.	System B1 (Bush 1952)	Alk. blue tetrazolium chloride	0.40	Anliker et al. (1961)
Desethyl-phosphamidon	Paper chromat.	System B1 (Bush 1952)	Alk. blue tetrazolium chloride	0.22	Anliker et al. (1961)
Chloroacetoacet-diethylamide	Paper chromat.	System B1 (Bush 1952)	Alk. blue tetrazolium chloride	0.61	Anliker et al. (1961)
Chloroacetoacet-ethylamide	Paper chromat.	System B1 (Bush 1952)	Alk. blue tetrazolium chloride	0.54	Anliker et al. (1961)
γ-Chloro-phosphamidon	Paper chromat.	System B1 (Bush 1952)	ChE inhib.	0.70	California Chem. Company (1963 c)
Phosphamidon	Paper chromat.[a]	Stat. phase: DMF in acetone. Mob. phase: hexane	Blue tetrazolium chloride	0.10	Bates (1965)
Phosphamidon	Paper chromat.[a]	Stat. phase: liq. paraffin in ether. Mob. phase: DMF and water	Blue tetrazolium chloride	0.92	Bates (1965)
Phosphamidon	TLC (silica gel)	Hexane-acetone (5+1)	Total P determ.	0.37	Abbott et al. (1967)
Phosphamidon	TLC (cellulose)	Stat. phase: 20% DMF in acetone. Mob. phase: methyl cyclohexane	Nitrobenzyl-pyridine	Not recorded	Ragab (1967)

Phosphamidon	TLC (silica gel) [b]	Dioxane	UV$_{254}$ and nitrobenzyl-pyridine	0.63	GUTH (1967)
Phosphamidon	TLC (silica gel) [b]	Acetonitrile	UV$_{254}$ and nitrobenzyl-pyridine	0.62	GUTH (1967)
Phosphamidon	TLC (silica gel) [b]	Acetone	UV$_{254}$ and nitrobenzyl-pyridine	0.60	GUTH (1967)
Phosphamidon	TLC (silica gel) [b]	Methylethyl ketone	UV$_{254}$ and nitrobenzyl-pyridine	0.51	GUTH (1967)
Phosphamidon	TLC (silica gel) [b]	Ethyl acetate	UV$_{254}$ and nitrobenzyl-pyridine		GUTH (1967)
Phosphamidon	TLC (silica gel) [b]	Ether	UV$_{254}$ and nitrobenzyl-pyridine	0.01	GUTH (1967)
Phosphamidon	TLC [c]	Cyclohexane-acetone-chloroform (70+25+5)	ChE inhib.	0.13 [d] 0.09 [e] 0.16 [f] 0.22 [g]	GETZ and WHEELER (1968)
Phosphamidon	TLC [c]	Acetone-isopropylether-cyclohexane (40+40+20)	ChE inhib.	0.67 [d] 0.40 [e] 1.00 [f] 0.74 [g]	GETZ and WHEELER (1968)
Phosphamidon	TLC (cellulose)	Stat. phase: 15% formamide in acetone. Mob. phase: benzene	Nitrobenzyl-pyridine and ChE inhib.	0.80	VOSS and GEISSBUEHLER (1967)
Desethyl-phosphamidon	TLC (cellulose)	Stat. phase: 15% formamidine in acetone. Mob. phase: benzene	UV$_{254}$, ChE inhib., GLC	0.55	Unpublished

[a] Reversed-phase system, two-dimensional. [b] Silica gel GF$_{254}$ (Merck), 0.3 mm. [c] Different adsorbents without binders. [d] Silic AR 4 (Mallinckrodt). [e] Silic AR 7 (Mallinckrodt). [f] Acid alumina (Woelm). [g] Florisil, 200 mesh and smaller (Matheson, Coleman, and Bell).

a) Bioassay for phosphamidon

The high toxicity of many insecticides, particularly organophosphates, toward *Daphnia magna* was utilized by Diemair and Knopf (1964) for the determination of organophosphates, including phosphamidon, in fruits and vegetables. The plant materials were extracted with water containing Tween 80. The limit of detection of the method was of the order of one p.p.m., which concentration killed 50 percent of the test organisms within five hours. The method is said to be useful for the detection of pesticide residues above tolerance level. Since organochlorine insecticides are much less toxic to *Daphnia* than organophosphates, this method has certain group specificity for organophosphates.

b) Cholinesterase inhibition residue analysis

Phosphamidon, desethylphosphamidon, and γ-chlorophosphamidon are potent inhibitors of cholinesterases, particularly plasma cholinesterases. This property was used for the first time for quantitative determinations of the compounds by the residue analysts of the former *California Chemical Company* (1963 a, b, and c), now *Chevron Chemical Company*. The basic procedure was later published by Pack et al. (1964). Human plasma cholinesterase activity measurements were carried out after a pre-inhibition period of one hour at 25° C. by a ΔpH-method using acetylcholine as a substrate. The limit of detection was reported to be 0.1 p.p.m. phosphamidon with a fair degree of accuracy, but it could be lowered to about 0.05 p.p.m. with an accuracy of ± 50 percent. Recoveries obtained with apples, fortified at levels between 0.05 and three p.p.m., were found to be between 60 and 90 percent (Pack et al. 1964). Cholinesterase inhibition residue analysis of desethylphosphamidon was carried out by the same procedure after separation of the metabolite from the parent compound on a cellulose column (*California Chemical Company* 1963 b).

γ-Chlorophosphamidon, a potent cholinesterase inhibitor, was separated from phosphamidon and desethylphosphamidon by a paper chromatographic method and semiquantitatively determined by a cholinesterase spray technique (*California Chemical Company* 1963 c). O-Desmethylphosphamidon was also determined by cholinesterase inhibition assay as phosphamidon, to which it had been converted by an esterification reaction with diazomethane (*CIBA.1964*).

A different approach for the determination of cholinesterase activities after inhibition by phosphamidon was chosen in the residue laboratories of *CIBA Ltd*. The chosen substrates were thiocholine esters, which upon enzymatic hydrolysis yield thiocholine, which reduces a sulfhydryl reagent, dithiobisnitrobenzoic acid, to the yellow anion of thionitrobenzoic acid, the absorbance of which is colorimetrically determined at 420 mμ (Ellman et al. 1961). For a detailed description of this procedure the reader is referred to section IV of the present article. The thiocholine ester method was also automated and extended by an additional thin-layer chromatographic step permitting the separation of phosphamidon from the closely related enolphosphates monocrotophos and dicrotophos (Voss and GEISS-BUEHLER 1967). The applied reversed-phase system, with formamide as the stationary and benzene as the mobile phase, also separated phosphamidon from desethylphosphamidon, and thus represented a further step toward a more specific residue method. The limit of detection of the automated cholinesterase inhibition method with human plasma cholinesterase was 0.05–0.1 p.p.m., but it can be increased by using peacock plasma cholinesterase which was found to be more sensitive to phosphamidon (Voss 1968). Recoveries with and without thin-layer chromatographic separation were higher than 80 percent. Other automated procedures based on the same analytical principle have recently been described by LEEGWATER and VAN GEND (1968) and by OTT (1968). The last-mentioned author ran an automated cholinesterase inhibition method simultaneously with an auto-mated total phosphate determination.

c) Diethylamine determination

The colorimetric procedure for quantitative determinations of OMPA (octamethyl pyrophosphoramide) (HALL et al. 1951) was modified by DITTMAN and his collaborators (cf. PACK et al. 1963) for residue analysis of phosphamidon. The method is based on the reaction of diethylamine with carbon disulfide and copper ion to form cupric diethyldithiocarbamate, which is water-insoluble and imparts a deep yellow color to certain organic solvents in which it is soluble. Since phosphamidon yields only one dialkyl-amine grouping upon hydrolysis instead of the four yielded by OMPA, the limit of detection is no better than one p.p.m. This method has a certain specificity, and does not detect desethylphosphamidon. Pesticides containing lower dialkylamino groups, such as dicrotophos and tetram, will interfere and certain fungicides, dithiocarbamates, and possibly some of the dialkyl carbamate pesticides will also give a positive reaction.

d) Total phosphate determination

Like many other organophosphates, phosphamidon can also be deter-mined by various modifications of the phosphomolybdenum blue method. This colorimetric procedure is preceded by the extraction of plant material, a suitable cleanup step, and the digestion of the organophosphate by oxidizing agents such as perchloric acid or sulfuric acid/hydrogen peroxide mixtures, and others. The procedure of ANLIKER and MENZER (1963) included a paper chromatographic cleanup but was of uncertain reliability for residues lower than 0.5 to one p.p.m. because of variable control values. A sensitivity of 0.1 p.p.m., however, was obtained with apple extracts purified on a column of activated charcoal (BREWERTON 1963). A further increase in sensitivity (0.04 p.p.m.) was achieved by cleaning up on a column of Celite and activated charcoal (STORHERR et al. 1964), before final separation of phosphamidon from fenitrothion by thin-layer chromatography *(CIBA* 1965). The phosphomolybdenum blue method used was that of Steller and Curry (1964). An automated total phosphate procedure for residue analyses of various organophosphates, including phosphamidon, was described by Ott and Gunther (1968) and Ott (1968).

e) Determination with nitrobenzylpyridine

A colorimetric determination based on the reaction between organophosphorus insecticides and 4-(*p*-nitrobenzyl) pyridine in slightly alkaline solution at 175° to 180° C. was described by Getz and Watts (1964). The sensitivity of this method was reported to be two µg. of organophosphate. Recovery values for phosphamidon have not been presented, but those obtained with other compounds varied between 60 and 112 percent for four different crops, at a residue level of one p.p.m. Extension of the procedure to crops such as kale, carrots, spinach, and lettuce, however, demonstrated the need for additional cleanup work. Non-specific color formation was also noticed in our own laboratories when traces of silica gel from TLC plates were present in the extracts, or when filter paper had been used for filtration of the extracts.

f) Double-paper chromatographic method

This method, which was described in detail by Pack *et al.* (1964), is based on the work of Anliker *et al.* (1961) of the *CIBA* laboratories. Methanol is used as an extractant, and the aqueous extract, obtained after solvent evaporation, is partitioned with hexane to remove interfering plant substances. Finally the aqueous phase is reextracted with methylene chloride, and the latter extract concentrated and subjected to a paper-chromatographic separation procedure (Bush 1952) for specific determination of phosphamidon, desethylphosphamidon, chloroacetoacetdiethylamide, and chloroacetoacetamide. When the developed chromatograms are dipped into an alkaline solution of blue tetrazolium chloride, blue spots appear, the color intensities of which are estimated visually and compared with known standards. The R_f-values for the four above-mentioned compounds are 0.32, 0.20, 0.59, and 0.49, respectively. Phosphamidon is detectable in amounts down to 0.2 µg., whereas desethylphosphamidon gives a measurable color in the chromatogram at levels of one µg. and above. Recoveries with various grops are 80 to 100%. The procedure is highly specific and very sensitive, with a limit of detection of 0.005 to 0.01 p.p.m. phosphamidon. Unfortunately, the double-paper chromatographic method is by far the most time-consuming residue method for phosphamidon, and as any official institutions no longer accept residue methods which are based on visual estimations it cannot be recommended in spite of its unique sensitivity and specificity.

g) Infrared spectroscopy

According to CROSBY and LAWS (1964) phosphamidon residues are detectable by infrared spectroscopy at concentrations down to approximately 0.1 p.p.m. The cleanup used by these authors was a combination of LAWS and WEBLEY's (1961) method (activated carbon — alumina chromatography) with a final gas chromatographic cleanup, during which the effluent was collected and then identified and measured by infrared spectroscopy. Recoveries slightly over 70 percent were obtained in an experiment with apples in the concentration range 0.2 to 4.0 p.p.m. The authors state that in many instances quantitative infrared measurements are tolerably close to parallel phosphorus determinations.

h) Gas chromatography

Like most other common organophosphates, phosphamidon can be analysed by gas chromatography. Two highly specific detectors particularly suitable for this purpose have been described, e. g., the phosphorus specific thermionic detector (GIUFFRIDA 1964) and the phosphorus-and-sulfur-specific Melpar flame photometric detector (BRODY and CHANEY 1966).

The insecticidal enolphosphates with their polar amido group require a suitable stationary phase on the column to achieve satisfactory retention times and separation. Various types have been recently described: Carbowax 20 M for the separation of monocrotophos and dicrotophos (BOWMAN and BEROZA 1967), a mixed-phase column of Apiezon L and Epicote 1001 for the separation of a number of different organophosphorus compounds, including the geometric isomers of phosphamidon (RUZICKA et al. 1967 a, and b), and a mixed-phase column of DC 200 and RF-1 (THORNTON and ANDERSON 1968). GIANG and BECKMAN (1968) separated dicrotophos from its metabolites by means of a short diethylene glycol adipate column. During development work on a new phosphamidon GLC-method at the CIBA residue laboratory the best separation effect was achieved on Carbowax 20 M stationary phases, but considerable adsorption caused difficulties in detecting nanogram amounts of phosphamidon and desethylphosphamidon. Less adsorption, however, was observed with a highly loaded SE 30 column, which was finally chosen for the method described under section IV b of the present chapter.

An electron capture GLC method (California Chemical Company 1963 d) for the two major nonphosphorylated metabolites, the N,N-diethyl- and the N-ethylamide of α-chloroacetoacetic acid, permits the determination of 0.05 and 0.025 p.p.m. of the monoamide and the diamide, respectively, if no natural plant substances are present with retention times similar to either or both of them.

IV. Recommended methods

We now describe two sensitive and relatively simple methods for analysis of phosphamidon residues. The first is based on human plasma cholinesterase inhibition and determination of residual enzyme activity using butyryl-thiocholine as a substrate and dithiobisnitrobenzoic acid as a reagent. If desethylphosphamidon and γ-chlorophosphamidon are not separated from the parent compound by suitable means (such as TLC) before enzyme inhibition assay, the method determines the total amount of all anticholinesterase compounds present in the plant extract. It thus accounts for all toxicologically important residues. The second is a newly developed GLC-method, using the Melpar flame photometric detector. The particular advantages of this latter method are a slightly higher sensitivity and a better specificity, since many insecticides interfering with the cholinesterase inhibition assay have retention times different from those of phosphamidon and desethylphosphamidon under the gas chromatographic conditions used.

a) Cholinesterase inhibition residue method

The plant material is macerated and extracted with chloroform, and the extract transferred into hexane by repeated evaporations. Phosphamidon is then extracted from hexane with water, and the aqueous phase used for cholinesterase (ChE) inhibition residue analysis.

1. Apparatus and reagents. — Vegetable macerator and mixer, rotating vacuum evaporator, centrifuge, temperature-constant water bath (30° C.), stopwatch, Klett-Summerson photoelectric colorimeter supplied with No. 42 filter, or equivalent, Celite 545 (*Johns-Manville Corp.*, N.Y.), organic solvents redistilled in glass or nanograde, and Soerensen phosphate buffer pH 7.0 and pH 8.0 (1/15 *M*).

The enzyme solution is prepared by diluting 0.5 ml. of outdated refrigerated human blood plasma with 100 to 150 ml. of buffer pH 8.0. The proper dilution factor must be determined for each new sample of pooled plasma.

Prepare the reagent stock solution by dissolving 100 mg. of 5,5′-dithiobis-2-nitrobenzoic acid (DTNB, *Aldrich Chemical Company*, Milwaukee, Wis., U.S.A.) in 50 ml. of buffer pH 7.0. Store in a refrigerator. This solution can be used for several weeks.

To prepare the substrate-reagent solution put 20 mg. of butyrylthiocholine iodide (BSCh, *Fluka AG.*, Buchs, Switzerland) into a 20-ml. volumetric flask, add two ml. of reagent stock solution, and make up to volume with distilled water. The substrate reagent solution should be kept in an ice bath during the day's work to minimize self-hydrolysis of BSCh.

The phosphamidon standard is prepared by dissolving 100 mg. of the technical grade compound in 100 ml. of acetone (= 1,000 p.p.m.). For recovery experiments and reference curves dilute with appropriate volumes of acetone or water, respectively.

2. Extraction and cleanup. — Chop the entire plant sample (500 to 1,000 g.) with a vegetable shredder. Place 25 g. of shredded plant material and 50 to 75 g. of anhydous sodium sulfate into the mixer and macerate at high speed for two to three minutes with 200 ml. of chloroform. Filter through *Schleicher and Schuell* "Sharskin" paper in a Büchner funnel, using vacuum. Disconnect vacuum, and repeat the extraction with a second portion of 150 ml. of chloroform. Filter again under reduced pressure, and finally rinse the filter cake with 50 ml. of chloroform. Concentrate the total chloroform extract to approximately ten ml. on the rotating evaporator (40° C. to 45° C.). Add 25 ml. of hexane, evaporate again to approximately five ml., and repeat this evaporation step twice more. Transfer the last concentrate to a ten-ml. graduated glass-stoppered test tube, rinse the evaporation flask with small portions of hexane, and make up to a volume of ten ml. Pipette a two-ml. portion into a 15-ml. centrifuge tube, add five ml. of distilled water and 30 to 40 mg. of Celite 545. Shake vigorously for at least 30 seconds, and centrifuge for ten minutes at 3,000 r.p.m. Remove the upper organic layer by suction and filter the aqueous phase through a folded paper. Then use a two-ml. aliquot of this solution for the cholinesterase assay.

   3. Reference inhibition curves. — Use a set of 11 special test tubes, of the type used as cells for measuring absorbance in the Klett-Summerson colorimeter. Pipette two ml. of distilled water into the tubes numbered one to three and two ml. of aqueous phosphamidon solutions (0.125, 0.25, 0.4, and 0.5 p.p.m., with duplicates) into the remaining eight tubes (Table IV,

Table IV. *Example for determining reference inhibition values for technical grade phosphamidon with human plasma cholinesterase*

Tube no.	Phosphamidon [a] (p.p.m.)	Absorbance		ChE activity (%)
		Individual	Mean	
1 (blank)	0	0	0	0
2, 3 (control)	0	0.480, 0.480	0.480	100
4, 5	0.125	0.360, 0.350	0.355	74
6, 7	0.25	0.280, 0.280	0.280	58
8, 9	0.40	0.192, 0.196	0.194	41
10, 11	0.50	0.154, 0.154	0.154	32

   [a] Concentration in test solution, two ml. of which is mixed with two ml. of enzyme solution.

Fig. 1). At the outset of the experiment start the stopwatch, add two ml. of buffer pH 8.0 into tube no. one and insert this tube into a temperature-constant water bath (30° C.). Thirty seconds later add two ml. of enzyme solution to tube no. two, place it in the water bath, and then proceed in the same way with the remaining tubes, always keeping the time interval of 30 seconds. After exactly 30 minutes (stopwatch time) add one ml. of substrate-reagent solution to tube no. one, shake, and put the tube back into the water bath. Do likewise with all the other tubes, still adhering to the time interval of 30 seconds. When the stopwatch shows 36 minutes, that is six minutes after the addition of the substrate-reagent solution[1], zero the colorimeter with tube no. one, and measure at intervals of 30 seconds the absorbance of the yellow solutions in the remaining ten tubes. If the timing is correct, all tubes are exposed for the same period to the particular conditions of pre-inhibition (30 minutes) and incubation with substrate (6 minutes). Calculate the percentages of residual enzyme activity

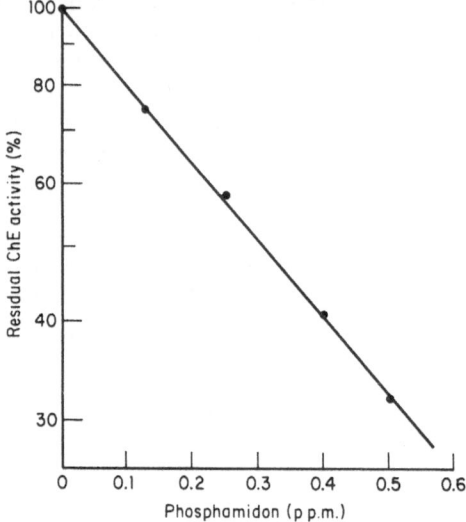

Fig. 1. Reference inhibition curve of technical grade phosphamidon obtained with human plasma cholinesterase (ChE)

in the four phosphamidon reference solutions by taking the non-inhibited control (no. two and three) as 100 percent active (Table IV), and plot these figures (log. scale) against inhibitor concentration (linear scale, Fig. 1). Check the validity of the reference curve frequently with two concentrations of phosphamidon, since aging of the plasma, even under frozen conditions of storage, can cause shifts in the reference curve. For each new sample of blood plasma prepare a new reference curve.

The slope of the reference curve is much affected by the percentage of γ-chlorophosphamidon present in the technical grade material because this impurity is a much stronger inhibitor of plasma cholinesterase than phosphamidon itself (Fig. 2). The curve presented in Fig. 1 is almost identical

---

[1] The incubation period of six minutes is not critical, but it must not exceed ten minutes. The only important point to keep in mind is that all samples must be incubated with the substrate for the same period. The Klett-Summerson colorimeter readings of tubes no. two and three should be of the order of 200 to 250, corresponding to absorbance values of 0.4 to 0.5.

with curve *B* in Fig. 2, which indicates that the technical grade sample used for establishing the first reference curve contained approximately two percent of the by-product. Since the rates of breakdown of phosphamidon and γ-chlorophosphamidon in living plants are much the same, we recommend the use of technical grade material for the preparation of reference curves. If, however, a thin-layer chromatographic cleanup is needed, which also separates the two compounds from each other, the analytical samples (thin-layer scrapings) are better analysed with pure phosphamidon as a reference.

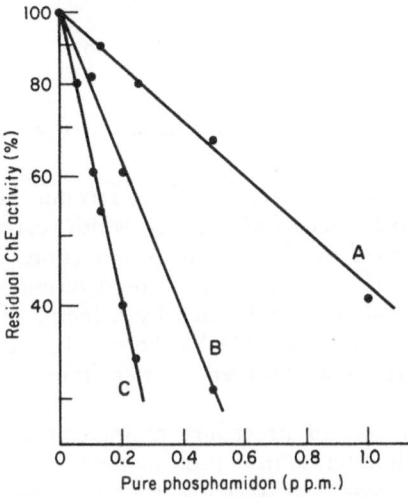

Fig. 2. Reference inhibition curves: *A* = purified phosphamidon, *B* = purified phosphamidon plus two percent γ-chlorophosphamidon with human plasma cholinesterase (ChE), and *C* = purified phosphamidon plus four percent γ-chlorophosphamidon with human plasma cholinesterase

4. Residue determinations. — After establishing the reference inhibition curve (Fig. 1), test two ml. of the final aqueous plant extracts as described above. The determined percentages of residual cholinesterase activity are then converted to the concentration of phosphamidon by using the reference curve. Since one ml. of the final extracts corresponds to one g. of plant material, no further residue calculations are necessary. The concentration of phosphamidon read from the abscissa of the reference curve equals the residue figure.

5. Limit of detection. — The limit of detection depends largely on the percentage of inhibition caused by extracts of untreated plants. Control inhibition values were determined for eight different crops in four to eight replicates (Table V). In all experiments inhibition was between ten and 20 percent which equals a residual cholinesterase activity of 80 to 90 percent.

Table V. *Control activity values obtained with extracts of untreated plant materials by the recommended cholinesterase inhibition residue method*

Crop	ChE activity (%)	Control value (p.p.m.) [a]	No. of replicates
Apples	88 ± 4	0.05 ± 0.02	8
Oranges (rind)	92 ± 6	0.04 ± 0.02	8
(flesh)	90 ± 4	0.05 ± 0.02	8
Strawberries	86 ± 3	0.06 ± 0.02	8
Cabbage	88 ± 4	0.05 ± 0.02	6
Carrots	89 ± 4	0.05 ± 0.02	8
Rice (grains)	92 ± 3	0.04 ± 0.02	4
Wheat (grains)	84 ± 3	0.07 ± 0.02	4

[a] Calculated on the basis of the reference curve presented in Figure 1.

For routine residue analysis we, therefore, recommend the establishment of a threshold activity value of 80 percent, which corresponds to a limit of detection of 0.1 p.p.m. If a higher limit of detection is required, the standard deviations of the control values are determined with a series of untreated samples, and then multiplied by a factor of 1.5 (FREHSE et al. 1962). With all crops tested (Table V) the standard deviation s was 0.02 p.p.m., which results in a so-called "lower limit of determination" of 0.03 p.p.m.

A more straightforward procedure to increase the sensitivity of the above described cholinesterase inhibition method is to use a more sensitive enzyme. Peacock plasma was recently found to contain ChE which is particularly sensitive to phosphamidon (Voss 1968). The limit of detection can be lowered by a factor of ten and, moreover, this enzyme permits the analysis of further diluted aqueous extracts of plant materials containing interfering substances (e. g., potatoes) while still retaining a relatively good sensitivity.

6. **Recoveries.** — The percentages of recovery obtained with the extraction and partitioning procedure described above are summarized in Table VI. Various crops, fortified with phosphamidon at levels from 0.04 to five p.p.m., were tested with both human and peacock plasma cholinesterase. In most experiments the recoveries were well above 80 percent, with standard deviations between five and ten percent.

7. **Specificity.** — The cholinesterase inhibition method described above is only semi-specific. Although a large number of non-polar insecticides do not interfere, because they remain in the hexane phase during the hexane-water partitioning cleanup, there are some polar compounds which pass into the aqueous extract together with phosphamidon (Table I). To render the method more specific, appropriate separation techniques have to be applied, of which TLC was found to be the most convenient. A chromatographic procedure, described by Voss and GEISSBUEHLER (1967), gave

Table VI. *Recovery values of phosphamidon with various types of crops*

Crop	Level of fortification [a]	Enzyme source [b]	No. of replicates	Recovered [c] (mean) (%)
Apples	1.2	ChE	8	85 ± 6
Oranges (rind)	0.8	ChE	8	83 ± 6
Strawberries	0.2	ChE	4	85 ± 9
	0.2	P-ChE	2	106
Cabbage	5.0	ChE	8	93 ± 8
Carrots	0.4	ChE	6	75 ± 5
Rice (grains)	0.08	P-ChE	2	89
	0.2	ChE	1	75
	0.8	ChE	2	85
Wheat (grains)	0.04	P-ChE	4	75 ± 7
	0.2	ChE	2	75
	0.2	P-ChE	4	94 ± 12
	0.8	ChE	2	82
Potatoes	0.2	P-ChE	4	80 ± 5
	0.8	P-ChE	4	83 ± 10

[a] Expressed as p.p.m.
[b] Human plasma cholinesterase (ChE) or peacock plasma cholinesterase (P-ChE).
[c] Control values (see Table V) subtracted.

satisfactory results even for the separation of phosphamidon from the closely related enolphosphates mono- and dicrotophos. The modified method is as follows:

Develop thin-layer plates (250-μ layer, MN 300 cellulose F 254, *Antec AG.*, Birsfelden, Switzerland) with freshly prepared 15 percent formamide *(Merck)* in acetone for approximately two hours. Briefly air-dry the plates and immediately use them for chromatography of the plant extract (when use of the plates is delayed separation effects and reproducibility will suffer). Apply the concentrated chloroform extract of the aqueous phase as a streak, and spot the reference compounds (50 to 100 μg.) at both margins of the plate. Develop in benzene, and locate the reference spots under the UV-lamp (254 mμ). Scrape off the corresponding zones of the chromatographed plant extract, and transfer the cellulose to a micro-chromatography column. Elute with water, and determine the amount of insecticide by cholinesterase inhibition (R_f-values = phosphamidon 0.80, dicrotophos 0.47, monocrotophos 0.15.) The same chromatographic method is also used for a separate determination of phosphamidon and desethyl-phosphamidon by either cholinesterase inhibition analysis or GLC. An example of TLC separation of the three enolphosphates mentioned above, and subsequent cholinesterase inhibition analysis (automated modification) is given in Fig. 4.

8. **Automated ChE inhibiton analysis.** — Since a variety of organo-phosphate and carbamate insecticides are analysed by the cholinesterase

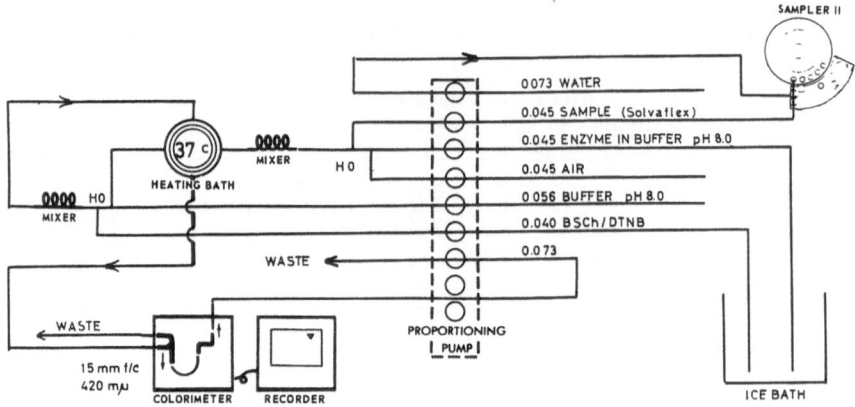

Fig. 3. Flow diagram of *Technicon* AutoAnalyzer® system for automated cholinesterase inhibition residue analysis of phosphamidon. Sample = aqueous solution of phosphamidon (references, aqueous extract of plant materials), Solvaflex tubing; ChE/buffer = one ml. of human plasma/200 to 250 ml. of buffer pH 8.0 or one ml. of peacock plasma/100 ml. of buffer pH 8.0; and BSCh/DTNB = 75 mg. of BSCh and five ml. of DTNB stock solution in buffer pH 7.0 made up to 100 ml. with distilled water

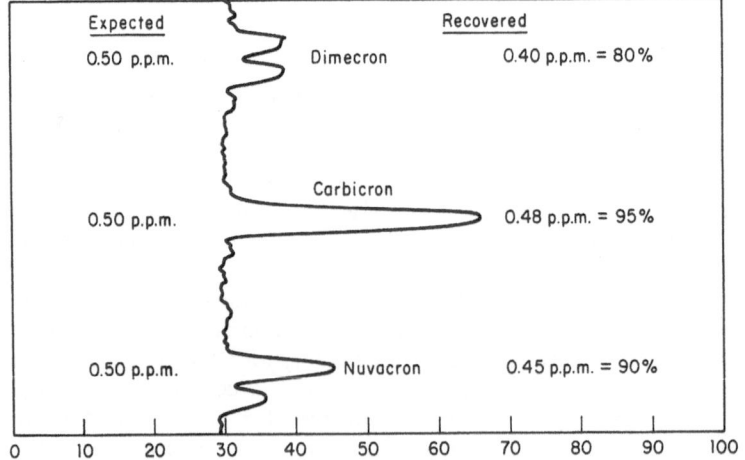

Fig. 4. AutoAnalyzer recordings of eluates of one cm. of TLC scrapings. Four g. of plant material fortified with 0.5 p.p.m. of each phosphate was applied to the TLC plate and chromatographed with benzene. Start = base margin of figure, solvent front = top margin of figure, Nuvacron = monocrotophos, Carbicron = dicrotophos, Dimecron = phosphamidon

inhibition technique in the *CIBA* laboratories, the manual procedure outlined above was automated with a view to obtaining better standardization of the test conditions and increasing the number of analyses carried out

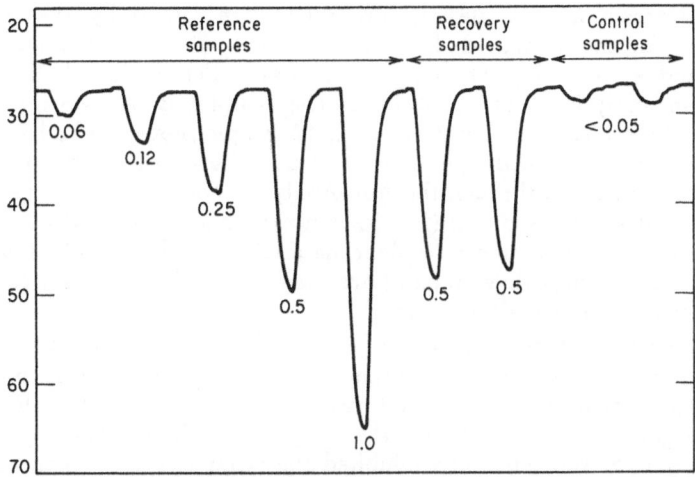

Fig. 5. Reference peaks of phosphamidon (0.06 to one p.p.m.), duplicate recovery peaks (tomatoes, 0.5 p.p.m., 90 and 86 percent recovery), and duplicate control values

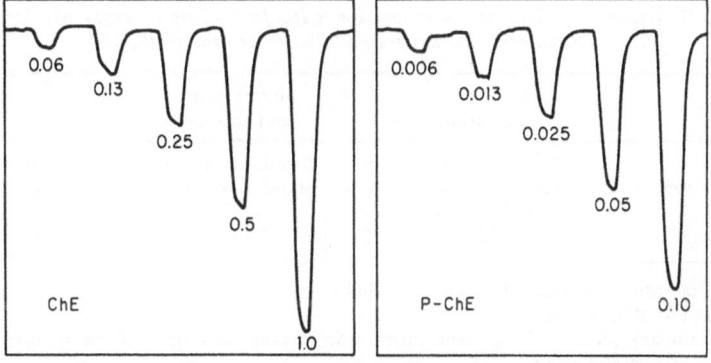

Fig. 6. Phosphamidon reference series recorded after automated inhibition of human plasma cholinesterase (ChE) and peacock plasma cholinesterase (P-ChE). Numbers indicate phosphamidon concentration in the test samples (p.p.m.)

during a day's work. The automated procedure was described in detail by Voss and Geissbuehler (1967) for inhibition residue analysis of enol-phosphates, including phosphamidon. Since some of the experimental conditions have been modified in the light of experience over the last two years, a brief description of the procedure will be given in the present chapter for the benefit of residue analysts who are familiar with the *Technicon* AutoAnalyzer® system [2].

_____

[2] Available from *Technicon Corporation,* Tarrytown, N.Y., U.S.A.

The extraction and cleanup procedures are essentially the same as described for the manual cholinesterase method. The flow diagram of the automated system (Fig. 3) gives all necessary details for cholinesterase inhibition analysis of phosphamidon. The sampler is best operated at a rate of 20 samples/hour, but 30 or even 40 samples/hour can be handled if necessary. For residue determinations pump a series of known phosphamidon concentration through the AutoAnalyzer, followed by the recovery, control, and residue samples (Fig. 5). Since with the standard extraction-partitioning procedure one g. of plant material results in one ml. of aqueous extract, the residue values (p.p.m.) are then obtained by a direct comparison of the references with the samples.

Two original AutoAnalyzer recordings of reference series over ranges of concentrations of 0.06 to one p.p.m. and 0.006 to 0.1 p.p.m. with human plasma and peacock plasma cholinesterase, respectively, are shown in Fig. 6. Again it was found that the sample of technical grade phosphamidon used for the references inhibited the latter enzyme ten times more than human plasma cholinesterase. Recovery values for four different crops are summarized in Table VII.

Table VII. *Recovery values of phosphamidon with four types of crops obtained by the automated cholinesterase inhibition residue method [a]*

Crop	Levels of fortification [b]	Extraction and cleanup	Recovered (%)
Carrots	0.4, 2.0, 10.0	Standard method	96 ± 8
Cauliflower	0.4, 2.0, 10.0	Standard method	84 ± 4
Celery	0.25, 0.50	TLC [c]	85 ± 8
Tomatoes	1.0	TLC [c]	92 ± 12

[a] Data from Voss and Geissbuehler (1967).
[b] Expressed as p.p.m.
[c] Stationary phase = 15 percent formamide in acetone, mobile phase = benzene; R_f-value of phosphamidon = 0.8.

b) Gas chromatographic residue method

Depending on the type of crop, the plant material is macerated and extracted with either acetonitrile or methylene chloride. After addition of water the organic solvent is evaporated, and phosphamidon and desethyl-phosphamidon are partitioned into chloroform. To remove oily materials from such materials as cotton seeds, the aqueous extract is washed with *n*-hexane before extraction with chloroform. The chloroform is replaced by *n*-hexane by repeated additions and evaporations of the latter solvent. The hexane solution is concentrated and applied to a silica gel column which is washed with an ethyl acetate-hexane mixture. Phosphamidon and desethylphosphamidon are eluted with ethyl acetate and separated by TLC

(Voss and Geissbuehler 1967). Quantitative determination is done by gas chromatography, using the Melpar flame photometric detector (Brody and Chaney 1966).

1. Apparatus. — Vegetable shredder (Bauknecht or Hobart type), grain- or seed-mill (e. g., a Condux Disc Mill), Omni-Mixer (or equivalent), rotating type evaporator, Kuderna-Danish evaporators kept in a water bath (80° to 90° C.) and equipped with three-ball Snyder columns, Soxhlet extraction apparatus equipped with 43 × 123 mm. thimbles, Chromato-charger of the Firmenich-type with a 100 µl. Hamilton syringe *(Camag, Muttenz, Switzerland)*, TLC-plates covered with MN 300 cellulose F_{254} *(Antec AG., Birsfelden, Switzerland)*, and chromatography columns, 400 × 15 mm. with coarse sintered disc G2.

2. Reagents. — Silica gel Woelm for adsorption chromatography of activity grade I *(M. Woelm, Eschwege, Germany)*. All organic solvents are redistilled in glass or are Nanograde. Phosphamidon and desethylphospha-midon standards are prepared by dissolving 100 mg. of the analytical grade compound in 100 ml. of benzene. For recovery experiments and reference curves dilute with appropriate volumes of *n*-hexane.

3. Extraction of vegetables and fruits, partitioning. — Place 50 g. of shredded vegetables or fruits into the jar of the Omni-Mixer and macerate with 100 ml. of acetonitrile at high speed for three minutes. Filter the resulting suspension through a 6-cm. Büchner funnel (layered with *Schlei-cher and Schuell* "Sharkskin" filter paper) by using slight vacuum. Rinse the jar and filter cake with two 25-ml. portions of acetonitrile. Combine the two filtrates in a round-bottom flask, add 40 ml. of distilled water and evaporate the acetonitrile using the rotating type evaporator (50° C.). Transfer the aqueous residue into a 100-ml. separating funnel and extract with three 50-ml. portions of chloroform, previously used for rinsing the evaporation flask. Collect the three chloroform fractions in a 500-ml. Erlenmeyer flask and remove traces of water by adding five g. of anhydrous sodium sulfate. Remove the sodium sulfate by filtration through a folded filter paper. Rinse the filter paper with 20 ml. of chloroform. Transfer the chloroform extract into a Kuderna-Danish evaporator and concentrate to about five ml. Remove chloroform by repeated additions of ten ml. of *n*-hexane (three times) and evaporation to three ml.

4. Extraction of oily crops, partitioning. — For residue analysis of oily crops, such as cotton seeds, place 50 g. of coarsely-milled plant material into the thimble of the Soxhlet apparatus. Extract with 350 ml. of methylene chloride for four hours using 500-ml. round-bottom flasks. Add 50 ml. of distilled water to the extract and remove methylene chloride by evaporation using the rotating evaporator (40° C.). Transfer the aqueous residue into a 500-ml. separating funnel, add 150 ml. of distilled water previously used for rinsing the evaporation flask, and extract with two 100-ml. portions of *n*-hexane. To avoid formation of emulsions shake carefully. Drain off the aqueous phase into another 500-ml. separating funnel, extract with three 100-ml. portions of chloroform, and collect the chloroform phases in a 500-ml. Erlenmeyer flask. Remove traces of water with anhydrous sodium sulfate, concentrate, and transfer the dissolved residue to hexane as described above.

5. Cleanup by column chromatography. — The use of a silica gel column for further cleanup is recommended for all plant materials mentioned above. The procedure is as follows: Fill the chromatographic column with *n*-hexane, adjust flow rate to one to two ml./min. and place two g. of the silica gel into the column. When the solvent head reaches the top of the silica gel close the stopcock and carefully pipet the hexane concentrate on top of the column. Rinse the tube of the Kuderna-Danish evaporator twice with three-ml. portions of 50 percent ethyl acetate in *n*-hexane, add the washings to the column, and wash the column with an additional portion (25 ml.) of the same organic solvent mixture. Elute phosphamidon and desethylphosphamidon with 50 ml. of ethyl acetate into a 100-ml. pear-shaped flask using the same flow rate. Reduce the volume of the eluate to five ml. using the rotating evaporator. Transfer the concentrate to a ten-ml. graduated tube, wash the evaporation flask with small portions of hexane, and fill up to a volume of ten ml. with hexane. This concentrate of which one ml. corresponds to five g. is used for GLC determination, if only phosphamidon is to be determined.

6. Separation of phosphamidon and deethylphosphamidon. — Concentrate the ethyl acetate eluate of the column to one ml. by using a gentle stream of air. Apply this concentrate as a ten-cm. streak to a cellulose plate, which has been pre-treated with a solution of 15 percent formamide in acetone, by means of the Chromatocharger. Spot references of phosphamidon and desethylphosphamidon (200 µg. of each compound) at both margins of the same plate, and develop with benzene for approximately 30 minutes. Locate the reference spots under the UV-lamp (254 mµ) and scrape off the corresponding zones from the central part of the plate. Elute each of the two scrapings with 30 ml. of acetone by using a ten-mm. microchromatography column, concentrate to a volume of ten ml., and use these solutions for GLC analysis of phosphamidon and desethylphosphamidon, respectively. One ml. of the extract corresponds to 5 g. of plant material.

7. Gas chromatography. — The gas chromatographic measurements were carried out with an F & M High Efficiency Gas Chromatograph (*Hewlett-Packard,* F & M Scientific Division, Avondale, Penna, U.S.A.) equipped with a Melpar flame photometric detector (*Tracor Analytical Instruments,* Austin, Texas, U.S.A.). The detector was furnished with the 526 mµ filter for the detection of phosphorus and was operated at 170° C. Columns were made of borosilicate glass, 600 × 3.2 mm. inner diameter, packed with 20 percent SE 30 silicone gum rubber on Gas Chrom Q (DMCS, 80/100 mesh). The column temperature was 160° C. for phosphamidon and 170° C. for desethylphosphamidon. Nitrogen was used as carrier gas at the rate of 230 ml./minute. An auxiliary supply of 20 ml. of nitrogen/minute was used between column and detector. Hydrogen, oxygen, and air flow were 160, 15, and 55 ml., respectively.

For reference curves prepare hexane solutions containing one to ten p.p.m. of phosphamidon and desethylphosphamidon, respectively. Inject quantities between one and 20 ng. in no more volume than five μl. Since the two isomers of phosphamidon are separated from each other under the GLC conditions applied, two peaks of different heights are obtained. Measure the peak heights of *cis*- and *trans*-phosphamidon, and that of desethylphosphamidon, and plot them against ng. insecticide on a log-log scale (Figs. 7 and 8). For determination of residues inject five μl. of the test solution, measure peaks heights, and compare with the standard curve.

8. Control values and recoveries. — Different control (untreated) samples of apples, brussels sprouts, orange pulp and rind, and cotton seeds were analysed by the procedure described above. When five μl. of the final concentrate were injected, none of the samples showed any significant peaks interfering with those of phosphamidon or desethylphosphamidon (Figs. 9 and 10). Under the gas chromatographic conditions described above, 0.5 to one ng. of the enolphosphates analysed showed reponses of one to two

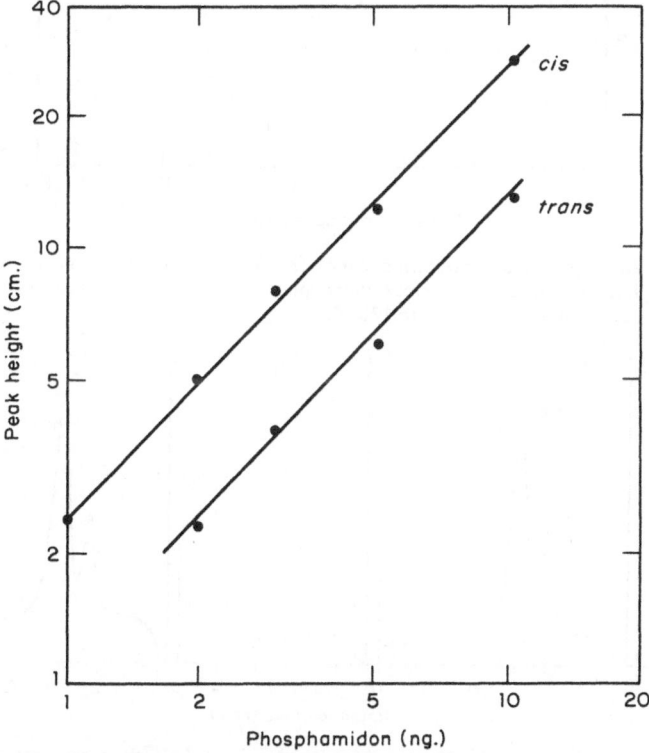

Fig. 7. Gas chromatographic calibration curves for *cis*- and *trans*-phosphamidon, obtained with the phosphorus-sensitive Melpar flame photometric detector; electrometer setting $10^3 \times 128$, column temperature 160° C.

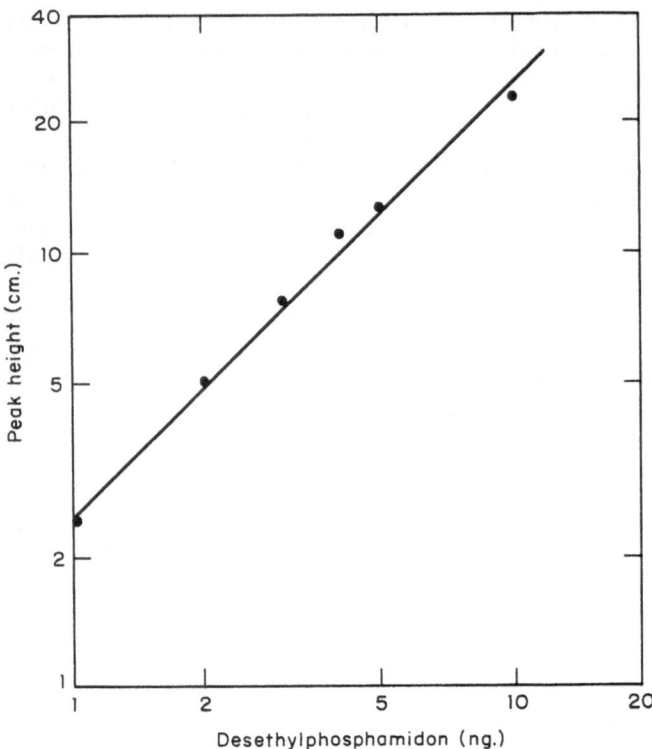

Fig. 8. Gas chromatographic calibration curve for desethylphosphamidon, obtained with the phosphorus-sensitive Melpar flame photometric detector; electrometer setting $10^3 \times 128$, column temperature 170° C.

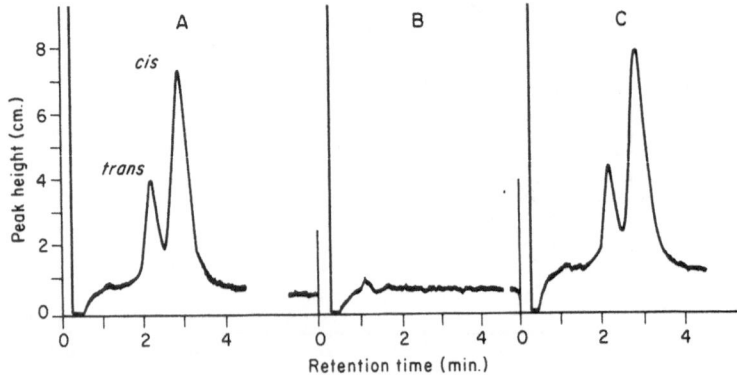

Fig. 9. Gas chromatograms of two ng. of phosphamidon (A), an extract of 25 mg. of untreated cotton seeds (B), and an extract of cotton seeds, fortified with phosphamidon before extraction at the 0.1 p.p.m. level (C); electrometer setting $10^3 \times 128$, column temperature 160° C.

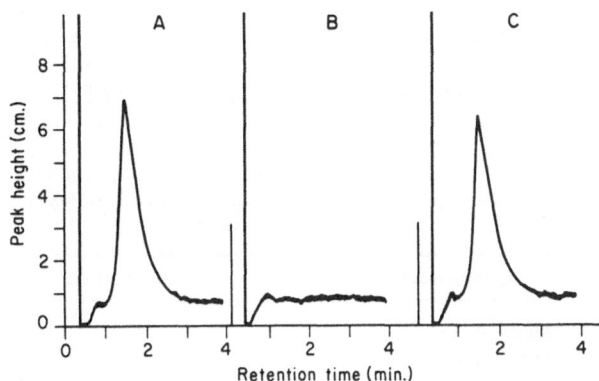

Fig. 10. Gas chromatograms of two ng. of desethylphosphamidon (*A*), an extract of 25 mg. of untreated cotton seeds (*B*), and an extract of cotton seeds, fortified with desethylphosphamidon before extraction at the 0.1 p.p.m. level (*C*); electrometer setting $10^3 \times 128$, column temperature 170° C.

cm. Peak heights in this range can be measured with reasonable accuracy. Since maximum quantities of control extracts injected were equivalent to 25 mg. of plant material, the limit of detection is 0.02 to 0.04 p.p.m.

Samples of different untreated crops were fortified with one and 0.1 p.p.m. enolphosphates and analysed by the described procedure. The recovery values obtained are given in Table VIII.

Table VIII. *Percentages of recovery and their standard deviations for phosphamidon and desethylphosphamidon in various plant materials, determined by the recommended GLC residue method*

Crop	Phosphamidon	Desethylphosphamidon
Apples	79 ± 10	95 ± 3
Oranges (rind)	87 ± 7	97 ± 12
Oranges (flesh)	81 ± 4	82 ± 4
Brussels sprouts	80 ± 12	89 ± 3
Cotton seeds	80 ± 4	88 ± 3

9. Specificity of the gas chromatographic method. — The hexane-water partitioning step used during the cleanup of plant extracts is the first efficient procedure to render the method specific. Only a limited number of well-known organophosphates, including phosphamidon, pass into the aqueous phase during partitioning (see Table I). All these compounds can be further separated from each other and identified by TLC and column chromatography. The separation of monocrotophos, dicrotophos, phosphamidon, and desethylphosphamidon by TLC has been described above. Furthermore, in contrast to phosphamidon and desethylphosphamidon, monocrotophos is not eluted from a silica gel column with ethyl acetate. Finally, the different retention times obtained under our GLC conditions allow a straightforward differentiation between phosphamidon and various related organophosphates of relatively high polarity (Table IX).

Table IX. *Relative retention times for some organophosphates which potentially interfere with the determination of phosphamidon and desethylphosphamidon in the recommended GLC residue method*

Organophosphate	Relative retention time
Mevinphos	0.15
Dicrotophos	0.52
Dimethoate	0.55
Monocrotophos	0.64
trans-Phosphamidon	0.74
Desethylphosphamidon	0.88
cis-Phosphamidon	1.00

V. Phosphamidon and the problem of multiple residue determinations

The government analyst and the control chemist in the food industry are always faced with the problem of multiple residue determinations of insecticides for quality assurance of edible crops with unknown spray history. Although the authors, as industrial chemists, have no practical experience of this type of analysis, they hope to make a useful contribution to the problem of phosphamidon analysis in the presence of multiple residues on the basis of some general methods described in the literature.

a) Extraction and cleanup

An extremely versatile extraction method, which will function with a large number of organophosphorus pesticides and plant materials, was described by Laws and Webley (1961). Its particular advantage is that it includes at the same time a relatively efficient column cleanup and a separation of the organophosphates into three major groups: water-soluble pesticides including phosphamidon (carbon column, chloroform), petroleum-soluble compounds (alumina column, light petroleum), and a third group of other organophosphates (alumina column, 15 percent ether-light petroleum mixture). This "general method" was used by Crosby and Laws (1964) prior to infrared spectroscopic identification of organophosphates, including phosphamidon, and also by Eder et al. (1964), who determined phosphamidon in the extract of water-soluble compounds by TLC. McKinley et al. (1964) tested this procedure among others and stated that it gave satisfactory results. Furthermore the extraction methods described in the present paper (see Table I) are suitable for multiple residue analysis provided the compounds present in the aqueous and hexane layers are separately determined by suitable detection methods. The hexane-water partitioning step represents at the same time an efficient cleanup procedure for the group of relatively polar compounds, including phosphamidon.

There is no general agreement about the types of extracting solvent to be used in a general extraction procedure. LAWS and WEBLEY (1961) recommended methylene chloride, GETZ (1962) suggested that acetonitrile could be used more widely since it readily extracts all organophosphates, and BATES (1965) in his general method selected acetone as an extracting solvent. From our own extended experience with phosphamidon it appears that all three solvents mentioned are suitable for extracting this particular insecticide. The choice of the extractant actually depends to a great extent on the type of crop under investigation.

b) Detection systems

Two different major systems for detecting a variety of organophosphates during multiple residue analysis have been successfully used in the past. Thin-layer chromatographic separation, with subsequent detection by suitable reagents, is still a widely used semiquantitative procedure in many control laboratories. A recently recommended spray reagent is 4-(p-nitrobenzyl)-pyridine, which reacts with both phosphates and thiophosphates (GETZ and WATTS 1964, GUTH 1967, RAGAB 1967). According to GUTH (1967) the smallest quantities of organophosphates detectable on thin-layer plates vary between 0.1 and one/μg. A second procedure, esterase inhibition after spraying the plates with blood plasma or liver homogenate preparations, is very sensitive to insecticidal phosphates including phosphamidon, whereas thiophosphates are only detected after activation to the corresponding phosphates. Various esterase-inhibition TLC procedures have been described by GETZ and FRIEDMAN (1963), ORTLOFF and FRANZ (1965), MENN and McBAIN (1966), ACKERMANN (1968), GETZ and WHEELER (1968), and MENDOZA et al. (1968). TLC is also useful for a separation of organophosphates prior to quantitative determinations by more sophisticated means.

A spray reagent specific for thiophosphates is potassium iodoplatinate (MacRAE and McKINLEY 1963, GUTH 1967), which of course will not detect phosphamidon residues. Its sensitivity is similar to that of nitrobenzyl pyridine.

The second and more important type of multiple residue detection systems is gas chromatography. Some of the publications on GLC methods available to the authors listed phosphamidon among a large number of other organophosphates (BACHE and LISK 1965, ABBOTT et al. 1965, RUZICKA et al. 1967 b). A more recent GLC method for simultaneous sensing of phosphorus- and sulfur-containing compounds by flame photometry (BOWMAN and BEROZA 1968) is a valuable addition to the older methods, since it differentiates between PO, PS, PS_2, and PS_3-organophosphates.

An approach based on determination of the ratio of different analytical responses (atoms or certain properties of the intact molecule) has also been proposed in connection with automated procedures. Otт (1968) proposed a dual AutoAnalyzer screening procedure for simultaneously measuring total phosphate and cholinesterase inhibition. Voss (1969) on the other hand, demonstrated that the differential inhibition of two different types of cholinesterase by organophosphates (and carbamates) can be used in an automated system as a screening method for anticholinesterase insecticides.

VI. Discussion

A number of methods for residue determination of phosphamidon were critically reviewed in the present chapter. Two procedures, cholinesterase inhibition analysis and gas chromatography, are recommended because of their sensitivity, ease of operation, and specificity.

The cholinesterase inhibition assay permits determination of residues down to a level of 0.01 p.p.m., provided a sensitive cholinesterase, such as that obtained from the peacock, is used. The particular advantage of this biochemical procedure is that it determines all cholinesterase-inhibiting toxic compounds present in plant material. The enzymatic procedure does not require a time-consuming and sophisticated cleanup, and is thus recommended for routine work, particularly when the automated procedure is used. The automated version of the thiocholine DTNB procedure gives better reproducibility, and practically no manual routine operations are necessary.

The GLC method is more specific, and is thus recommended for crops containing multiple residues of relatively polar organophosphates. The different retention times of phosphamidon and other related compounds allow a straightforward identification. The GLC procedure even allows the differentiation of cis- from trans-phosphamidon. Because of naturally occurring anticholinesterase plant constituents, the cholinesterase inhibition method is not applicable to certain crops such as potatoes, tobacco, and hops, or to alfalfa flour and certain oily seeds without a complicated cleanup. If, however, a thorough cleanup is needed because of high concentrations of interfering plant substances, or if a particularly high sensitivity is required, the authors recommend the GLC procedure because it gives specific and informative results.

Summary

The present chapter summarizes and reviews the different methods which have been used for residue determinations of phosphamidon, its major metabolite desethylphosphamidon, and γ-chlorophosphamidon in plant materials. The methods cover a wide range of possibilities, such as bioassays, biochemical tests, purely analytical determinations, and modern physico-chemical procedures. Out of a total of eight procedures reviewed, two methods were particularly recommended because of the sensitivity, ease of operation, and codetermination of toxic metabolites on one hand (cholinesterase inhibition analysis) and specificity on the other hand (GLC method). A detailed description of extraction, cleanup, and final determination is given for the two recommended methods. Also, phosphamidon determinations in connection with multiple residue analysis for quality assurance of food have been briefly discussed.

Résumé *

Méthodes de détermination des résidus de phosphamidon

Ce chapitre donne une vue d'ensemble des différentes méthodes qui ont servi à la détermination, dans du matériel végétal, des résidus du phosphamidon de son métabolite principal, le deséthylphosphamidon, ainsi que du γ-chlorophosphamidon. Ces méthodes comprennent des dosages biologiques et biochimiques, des déterminations purement analytiques et des procédés physico-chimiques modernes. Deux des huit procédés cités sont particulièrement recommandés pour la détermination des résidus du phosphamidon: la méthode enzymatique qui est basée sur l'inhibition de la cholinestérase est remarquable par sa sensibilité, sa facilité opératoire et son efficacité qui s'étend aux métabolites toxiques. Le procédé de chromatographie en phase gazeuse se distingue par sa spécificité. Une description détaillée de l'extraction, de la purification et de la détermination finale est donnée pour les deux méthodes recommandées. Finalement une brève discussion porte sur la détermination du phosphamidon en présence d'autres pesticides, problème qui se pose surtout lors du contrôle des aliments par les laboratoires officiels.

Zusammenfassung **

Rückstandsanalytische Methoden für Phosphamidon

Das vorliegende Kapitel gibt einen Überblick über verschiedene Methoden, die zur Bestimmung von Rückständen von Phosphamidon, dessen Hauptmetaboliten Desäthylphosphamidon, sowie γ-Chlorphosphamidon in Pflanzenmaterial verwendet worden sind. Unter diesen Methoden finden sich sowohl biologische und biochemische Tests, als auch rein analytische und physikalisch-chemische Verfahren. Zwei von insgesamt acht erwähnten Methoden wurden für Phosphamidon-Rückstandsanalysen besonders empfohlen: Das Cholinesterasehemmverfahren zeichnet sich durch hohe Empfindlichkeit, Einfachheit und die Miterfassung toxischer Metaboliten aus; bei der gaschromatographischen Methode dominiert die Spezifität. Für beide Verfahren wurden Extraktion, Reinigung und Endbestimmung ausführlich beschrieben. Die Bestimmung von Phosphamidon in Gegenwart anderer Pflanzenschutzmittel, ein Problem der Qualitätskontrolle, wurde abschließend erörtert.

---

* Traduit par les auteurs.
** Übersetzt von den Autoren.

References

Abbott, D. C., N. T. Crosby, and J. Thomson: The use of thin-layer and semi-preparative gas-liquid chromatography in the detection, determination, and identification of organo-phosphorus pesticide residues. In: P. W. Skallis, ed. Proc. SAC Conf., Nottingham, Eng., p. 121. Cambridge: Heffer and Sons (1965).

—, A. S. Burridge, J. Thomson, and K. S. Webb: A thin-layer chromatographic screening test for organophosphorus pesticide residues. Analyst 92, 170 (1967).

Ackermann, H.: Dünnschichtchromatographisch-enzymatischer Nachweis phosphororganischer Insektizide: Aktivierung schwacher Esterasehemmer. J. Chromatog. 36, 309 (1968).

Anliker, R., and R. E. Menzer: Method for phosphamidon residue analysis. J. Agr. Food Chem. 11, 391 (1963).

—, E. Beriger, M. Geiger, and K. Schmid: Über die Synthese von Phosphamidon und seinen Abbau in Pflanzen. Helv. Chim. Acta 44, 1622 (1961).

Bache, C. A., and D. J. Lisk: Determination of organophosphorus insecticide residue using the emission spectrometric detector. Anal. Chem. 37, 1477 (1965).

Bates, J. A. R.: A general method for the determination of organo-phosphorus pesticide residues in foodstuffs. Analyst 90, 453 (1965).

Bowman, M. C., and M. Beroza: Gaschromatographic analysis of 3-hydroxy-N-methyl-cis-crotonamide-dimethylphosphate (Azodrin) and 3-hydroxy-N,N-dimethyl-cis-crotonamide dimethylphosphate. J. Agr. Food Chem. 15, 465 (1967).

— — Gaschromatographic detector for simultaneous sensing of phosphorus and sulfur containing compounds by flame photometry. Anal. Chem. 40, 1448 (1968).

Brewerton, H. V.: Phosphamidon residues in apples. New Zealand J. Sci. 6, 259 (1963).

Brody, S., and J. E. Chaney: Flame photometric detector. The application of a specific detector for phosphorus and sulfur compounds. J. Gas Chromatog. 4, 42 (1966).

Bush, I. E.: Methods of paper chromatography of steroids applicable to the study of steroids in mammalian blood and tissues. Biochem. J. 50, 370 (1952).

California Chemical Co. (now Chevron Chemical Co.), Ortho Division, Research and Development, Richmond, Calif.: Phosphamidon. Preliminary report on the isolation and properties of α- and β-isomers. Unpublished report (1962).

— Analysis of phosphamidon residues. Method RM-4. Unpublished report (1963 a).

— Chromatographic separation and analysis of phosphamidon and desethylphosphamidon residues. Method RM-4 A. Unpublished report (1963 b).

— Paper chromatographic method for the estimation of residues of gamma-chlorophosphamidon. Method RM-4 C. Unpublished report (1963 c).

— Electron capture GLC method for the residue analysis of two non-phosphorylated phosphamidon metabolites. Method RM-4 B. Unpublished report (1963 d).

CIBA Ltd., Basle, Switzerland: Method for desmethylphosphamidon residue analysis. Unpublished report (1964).

— Fenitrothion/Phosphamidon. Residue analysis in plant material. Unpublished report (1965).

Crosby, N. T., and E. R. Laws: The use of infrared spectroscopy in the analysis of pesticide residues. Analyst 89, 319 (1964).

Diemair, W., and K. Knopf: Über die Eignung von Daphnia magna zur Ermittlung von Rückständen auf frischem Obst und Gemüse. Zeitschr. analyt. Chem. 200, 226 (1964).

Eder, F., H. Schoch, and R. Mueller: Nachweis von Insektizidrückständen (chlorierte Kohlenwasserstoffe, Phosphorsäureester) in bzw. auf Obst und Gemüse mit Hilfe der Papier- und Dünnschichtchromatographie. Mitt. Lebensmitteluntersuchung u. Hygiene 55, 98 (1964).

Ellmann, G., D. Courtney, V. Andres, and R. M. Featherstone: A new and rapid colorimetric determination of acetylcholinesterase activity. Biochem. Pharmacol. 7, 88 (1961).

FREHSE, H., H. NIESSEN, and H. TIETZ: Zur Anwendung des p.p.m.-Begriffs im Bereich sehr kleiner Pflanzenschutzmittel-Rückstände. Pflanzenschutznachrichten Bayer 15, 113 (1962).

GETZ, M. E.: Six phosphate pesticide residues in green leafy vegetables. Cleanup method and paper chromatographic identification. J. Assoc. Official Agr. Chemists 45, 393 (1962).

—, and S. J. FRIEDMANN: Organophosphate pesticide residues: A spot test for detecting cholinesterase inhibitors. J. Assoc. Official Agr. Chemists 46, 707 (1963).

—, and R. R. WATTS: Application of 4-(p-nitrobenzyl) pyridine as a rapid quantitative reagent for organophosphate pesticides. J. Assoc. Official Agr. Chemists 47, 1094 (1964).

—, and H. G. WHEELER: Thin-layer chromatography of organophosphorus insecticides with several adsorbents and ternary solvent systems. J. Assoc. Official Agr. Chemists 51, 1101 (1968).

GIANG, B. Y., and H. F. BECKMAN: Determination of Bidrin, Azodrin, and their metabolites with the thermionic detector. J. Agr. Food Chem. 16, 899 (1968).

GIUFFRIDA, L.: A flame ionisation detector highly selective and sensitive to phosphorus — A sodium thermionic detector. J. Assoc. Official Agr. Chemists 47, 293 (1964).

GUTH, J. A.: Ein dünnschichtchromatographischer Trennungsgang für insektizid wirksame Phosphorsäureester. Pflanzenschutzberichte (Wien) 45, 129 (1967).

HALL, S. A., J. W. STROHLMAN, and M. S. SCHECHTER: Colorimetric determination of octamethyl-pyrophosphoramide. Anal. Chem. 23, 1866 (1951).

LAWS, E. Q., and D. J. WEBLEY: The determination of organophosphorus insecticides in vegetables: A general method for insecticide residues. Analyst 86, 249 (1961).

LEEGWATER, D. C., and H. W. VAN GEND: Automated differential screening method for organophosphorus pesticides. J. Sci. Food Agr. 19, 513 (1968).

MACRAE, H. F., and W. P. MCKINLEY: Chromatographic identification of some organophosphate insecticides in the presence of plant extracts. J. Agr. Food Chem. 11, 174 (1963).

MCKINLEY, W. P., D. E. COFFIN, and K. A. MCCULLY: Cleanup processes for pesticide residue analysis. J. Assoc. Official Agr. Chemists 47, 863 (1964).

MENDOZA, C. E., P. J. WALES, H. A. MCLEOD, and W. P. KINLEY: Enzymatic detection of ten organophosphorus insecticides and carbaryl on thin-layer chromatograms: An evaluation of indoxyl, substituted indoxyl and l-naphthyl acetates as substrates of esterase. Analyst 93, 34 (1968).

MENN, J. J., and J. B. MCBAIN: Detection of cholinesterase-inhibiting insecticide chemicals and pharmaceutical alkaloids on thin-layer chromatograms. Nature 209, 1351 (1966).

ORTLOFF, R., and P. FRANZ: Zwei neue Methoden der biochemischen Lokalisierung von phosphorhaltigen Insektiziden auf Dünnschichtchromatogrammen. Zeitschr. Chem. 5, 388 (1965).

OTT, D. E.: Dual simultaneous AutoAnalyzer for screening some insecticide residues. A total phosphorus system and a new anticholinesterase system. J. Agr. Food. Chem. 16, 874 (1968).

—, and F. A. GUNTHER: Automated analysis of organophosphorus insecticides by wet digestion-oxidation and colorimetric determination of the derived orthophosphate. J. Assoc. Official Agr. Chemists 51, 697 (1968).

PACK, D. E., J. N. OSPENSON, and G. K. KOHN: Phosphamidon — In G. Zweig, ed.: Analytical methods for pesticides, plant growth regulators, and food additives, vol. 2, p. 375. New York-London: Academic Press (1964).

RAGAB, M. T. H.: 4-(p-nitrobenzyl) pyridine as a spray reagent for organophosphorus pesticides and some of their breakdown products on thin-layer chromatograms. Bull. Environ. Contamination Toxicol. 2, 279 (1967).

RUZICKA, J. H., J. THOMSON, and B. B. WHEALS: The gaschromatographic examination of organophosphorus pesticides and their oxidation products. J. Chromatog. 30, 92 (1967 a).

— — — The gaschromatographic determination of organophosphorus pesticides. Part II. A comparative study of hydrolysis rates. J. Chromatog. 31, 37 (1967 b).

Storherr, R. W., M. E. Getz, R. R. Watts, S. J. Friedmann, F. Erwin, L. Giuffrida, and F. Ives: Identification and analysis of five organophosphate pesticides. Recoveries from crops fortified at different levels. J. Assoc. Official Agr. Chemists 47, 1087 (1964).

Steller, W. A., and A. N. Curry: Measurement of residues of Cygon and its oxygen analog by total phosphorus determination after isolation by thin-layer chromatography. J. Assoc. Official Agr. Chemists 47, 645 (1964).

Thornton, J. S., and C. A. Anderson: Determination of residues of Di-Syston and metabolites by thermionic emission flame gas chromatography. J. Agr. Food Chem. 16, 895 (1968).

Voss, G.: Peacock plasma, a useful cholinesterase source for inhibition residue analysis of insecticidal carbamates. Bull. Environ. Contamination Toxicol. 3, 339 (1968).

— AutoAnalyzer of insecticidal organophosphates and carbamates, based on cholinesterase inhibition. In press: Technicon International Congress, Chicago, June 4–6 (1969).

—, and H. Geissbuehler: Automated residue determinations of insecticidal enolphosphates. Mededelingen Rijksfaculteit Landbouwwetenschappen Gent 32, 877 (1967).

Chapter 8

Rate of degradation of phosphamidon and residue values

By

G. Voss and H. Geissbühler

Contents

I. Introduction

Phosphamidon, a systemic insecticide, is widely used for the control of phytophagous insects and mites on a great variety of cultivated plants. In consequence, many food crops have been analysed during the past ten years for residues of phosphamidon and its major metabolites. The aim of the present chapter is to summarize these data and to present a fairly comprehensive picture of the rates of degradation in various crops, and of phosphamidon residues present in them at the time of harvest. For a more complete understanding of the residue situation the reader is referred to the chapters on the toxicology, behavior, and metabolism of phosphamidon in plants and animals.

II. Rate of degradation in plants

The present chapter is concerned with the rates of degradation of phosphamidon, desethylphosphamidon, and γ-chlorophosphamidon in living plants. Experiments to determine half-lives were carried out in the labo-

133

ratory or greenhouse. In field experiments the concentrations used were within the ranges recommended for insect control.

a) Phosphamidon

Phosphamidon is quite rapidly metabolized to the biologically active intermediate desethylphosphamidon and various non-toxic substances. Details of the metabolic pathways are given in the chapter on phosphamidon metabolism. The rate of breakdown in plants is of interest to residue chemists and toxicologists and has been studied in various plant species. Some typical breakdown curves for leafy crops are presented in Figure 1, from which it is clear that the half-life of phosphamidon on green metabolizing leaves is of the order of one to 1.5 days. Thus, residues remaining in plant material after normal application generally decrease to less than one p.p.m. within the first seven to ten days after application.

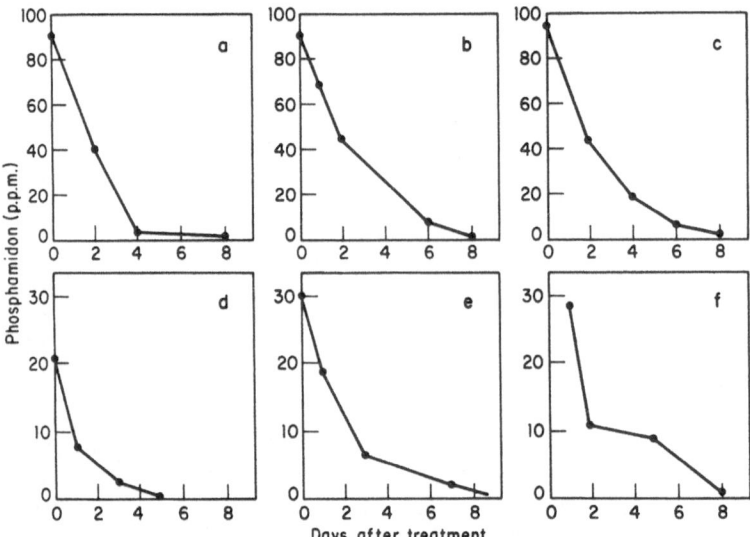

Fig. 1. Phosphamidon degradation curves obtaines with various leafy crops: $a = bean$ *plants,* greenhouse (25° C.), sprayed to run-off with 0.02 percent a. i., residue analysis by automated cholinesterase inhibition method (*CIBA* 1968 a); $b = rice$ *foliage,* greenhouse, ultra-low volume application, residue analysis by automated cholinesterase inhibition method (*CIBA* 1968 b); $c = potato$ *foliage,* growth chamber (24° C.), plants dipped into phosphamidon solution, residue analysis by total phosphate determination (Menzer and Ditman 1963 a); $d = spinach,$ sprayed in the field with 0.12 percent a. i., residue analysis by double paper chromatography *(California Chemical Company* report CSC 513/214–25, 1960); $e = rice$ *foliage,* 0.45 kg. a. i./ha., residue analysis by cholinesterase inhibition method *(CIBA-Florida,* research report CF-676, 1966); and $f = Swiss$ *chard,* sprayed in the field with 0.04 percent a. i., residue analysis by total phosphate determination *(CIBA* 1964 b)

The effect of environmental factors on the degradation of phosphamidon was studied by MENZER and DITMAN (1963 a). When snap beans
were treated with 0.5 pint of phosphamidon/50 gallons of water, residues
decreased from approximately 100 p.p.m. to three p.p.m. within ten days
at 16° C. At 24° C. a residue level of three p.p.m. was reached after eight
days and at 32° C. it took six days for phosphamidon to degrade from the
original 100 p.p.m. to three p.p.m. Thus, a temperature increase of 15° C.
to 20° C. doubles the rate of breakdown. Similar results were obtained
with potato foliage. The same authors also investigated the effect of day
length on phosphamidon degradation and found that it was not pronounced. Only with a very short day (one hour) was the rate of breakdown definitely reduced. The degradation of phosphamidon was also affected by the concentration of the insecticide applied to the plants. With
very high initial residues (500 to 2,000 p.p.m.) the slopes of the breakdown curves were less steep than those obtained with initial residues of
approximately 100 p.p.m.

The rate of disappearence of phosphamidon from fruit crops is considerably less rapid than in several leafy crops. Extensive studies of phosphamidon residues in apples over a period of almost two months have
been described by BREWERTON (1963). The initial residues were of the order
of 1.0 to 1.2 p.p.m., decreasing to 0.2 p.p.m. over a period of 50 days. The
half-life was estimated by BREWERTON (1963) to be approximately 25 days.
Although a half-life of this length has not been confirmed by other analysts,
it is evident from the results of MENZER and DITMAN (1963 b) and from
those obtained in the residue laboratories of the former *California Chemical Company* and *CIBA* that phosphamidon breakdown on various fruits
is slow when compared with that in metabolizing plant leaves.

b) Desethylphosphamidon and γ-chlorophosphamidon

The rate of degradation of desethylphosphamidon, the toxic plant
metabolite of phosphamidon, is similar to that of the parent compound.
ANLIKER et al. (1961) and MENZER and DITMAN (1963 a) found the N-des-
ethylated compound to be slightly less stable than phosphamidon itself
when they followed the appearance and disappearance of the metabolite
in green plants. However, a direct comparison of the rates of breakdown
of the two substances is difficult because all the desethylphosphamidon is
within the plant, and exposed to further metabolism, whereas the residues
of the parent compound are partly on the surface of the plant and partly
within it, both fractions being determined.

In a recent experiment the relative rates of breakdown of phosphamidon, desethylphosphamidon, and γ-chlorophosphamidon were determined
after direct application of these compounds to bean plants by dipping the
leaves (Fig. 2). In this experiment the desethyl-compound was slightly
more stable than phosphamidon and γ-chlorophosphamidon. The small difference, however, is not significant for practical purposes, so that for

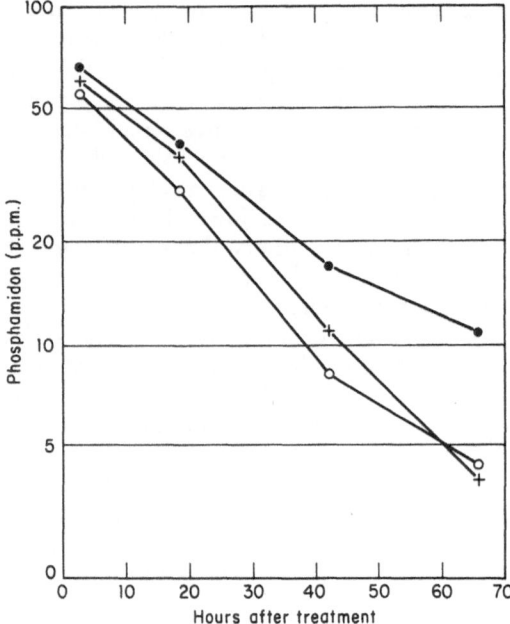

Fig. 2. Degradation of phosphamidon (+), desethylphosphamidon (•), and γ-chloro-phosphamidon (o). Leaves of bean plants were dipped into a solution of 0.05 percent a. i. and analysed at various time intervals by cholinesterase inhibition. During the experiment plants were kept under the following conditions: 14-hour day (25° C.), 10-hour night (18° C.), 75 percent, r. h., light intensity 10,000 Lux

residue analyses the rates of degradation of all three compounds can be assumed to be equal. This is particularly important in the case of residue methods based on cholinesterase inhibition, since the highly active inhibitor γ-chlorophosphamidon would give rise to considerable error in the analytical determination of phosphamidon if its rate of breakdown differed greatly from that of the parent compound.

III. Residues in plants

Most of the residue values of phosphamidon and its metabolites in plants have been obtained by the Ortho Division of the former *California Chemical Company,* now *Chevron Chemical Company,* and by the residue laboratory of *CIBA* Ltd., Basle, Switzerland. Some data taken from published papers or reports of research institutes are mentioned specifically.

a) Phosphamidon

Residue levels determined after the treatment of crops with phosphamidon in various climates and at various rates of application are listed in

Table I. *Residues of phosphamidon in fruit crops, expressed as p.p.m. after the indicated time intervals*

a) *Apples:* U.S.A. and Switzerland (cholinesterase inhibition or double paper chromatography)

Type of sample	Dosage (kg. a.i./ ha.)	Conc. (%)	No. appl.	Phosphamidon (p.p.m.)						
				Days after application						
				0	1	3–4	6–8	9–12	14–18	20–25
Fruit	1.4	0.06	1	—	2.8	1.6	< 0.1	< 0.1	< 0.1	< 0.1
	1.1	dust	7	—	2.0	1.0	< 0.1	< 0.1	< 0.1	—
	0.8	0.03	4	2.0	1.8	1.4	0.6	0.5	0.3	0.2
	0.5	0.03	5	—	—	—	—	—	< 0.1	0.2
	2.2	0.03	4	—	—	—	—	—	0.6	0.3
	1.4	0.03	4	—	—	—	—	—	1.0	0.8
	1.1	0.03	4	—	—	—	—	—	2.2	1.8
Fruit, young	—	0.02	1	7.0	9.9	3.9	1.4	—	1.1	0.4
	—	0.04	1	9.6	8.4	0.8	0.8	—	1.2	0.1
	—	0.04	1	1.0	—	—	0.5	—	0.2	0.12

b) *Blueberries:* Canada (cholinesterase inhibition)

Type of sample	Dosage (kg. a.i./ha.)	Conc. (%)	No. appl.	Phosphamidon (p.p.m.)		
				Days after application		
				2	3	9
Fruit	0.2	—	1	0.2	< 0.1	< 0.1

c) *Cantaloupes:* U.S.A. (phosphate determination)

Type of sample	Conc. (%)	No. appl.	Phosphamidon (p.p.m.)			
			Days after application			
			0	1	2	3
Fruit	0.06	4	< 0.1	< 0.1	< 0.1	< 0.1
	0.03	5	< 0.1	—	—	< 0.1

d) *Cherries:* Switzerland (double paper chromatography)

Type of sample	Conc. (%)	No. appl.	Phosphamidon (p.p.m.)				
			Days after application				
			1	4	8	13	15
Fruit	0.04	1	7.9	4.2	1.0	0.9	0.3

Table I (continued)

e) *Cranberries:* U.S.A. (diethylamine determination)

Type of sample	Dosage (kg. a.i./ha.)	No. appl.	Phosphamidon (p.p.m.)	
			Days after application	
			0	7
Fruit	0.55	1	3.4	1.5

f) *Grapes:* U.S.A. (cholinesterase inhibition)

Type of sample	Conc. (%)	No. appl.	Phosphamidon (p.p.m.)			
			Days after application			
			1	6	11	21
Fruit	0.06	3	2.7	1.4	0.4	0.2
	0.12	3	4.8	2.8	0.6	0.3

g) *Grapefruit:* U.S.A. (cholinesterase inhibition)

Type of sample	Dosage (kg. a.i./ha.)	Conc. (%)	No. appl.	Phosphamidon (p.p.m.)	
				Days after application	
				1	15
Fruit	2.8	0.15	4	—	< 0.1
	3	0.19	4	—	< 0.1
	—	0.12	—	2.2	—
Fruit, washed	—	0.12	—	0.13	—
Molasses	—	0.12	—	< 0.1	—

h) *Lemons:* U.S.A. (cholinesterase inhibition)

Type of sample	Dosage (kg. a.i./ha.)	Conc. (%)	No. appl.	Phosphamidon (p.p.m.)
				Days after application
				15
Fruit	2.8	0.15	4	< 0.1
	3	0.19	4	< 0.1

Table I (continued)

i) *Oranges:* U.S.A. (cholinesterase inhibition)

Type of sample	Dosage (kg. a.i./ha.)	Conc. (%)	No. appl.	Phosphamidon (p.p.m.)				
				Days after application				
				0	4–5	10–12	15–18	21–26
Fruit	5.5	0.06	4	4.1	2.5	0.6	—	0.33
	6	0.4	4	—	—	—	0.3	—
	6	0.4	4	—	—	—	0.4	—
	6	0.4	1	—	—	—	0.2	—
	6	0.4	1	—	—	—	0.15	—
	2.8	0.3	1	—	—	—	<0.1	—
Juice	4.2	0.03	1	0.3	<0.1	<0.1	<0.1	—
	8.4	0.06	1	0.3	0.2	0.2	0.2	—
	16.8	0.12	1	1.3	0.4	0.2	<0.1	—
Pulp	4.2	0.03	1	0.9	0.6	0.4	<0.1	—
	8.4	0.06	1	2.6	1.6	0.5	0.3	—
	16.8	0.12	1	3.8	3.0	1.4	0.4	—
Fruit, young	4.8	0.03	1	1.1	0.8	<0.1	0.2	<0.1
	9.6	0.06	1	1.6	1.3	0.3	<0.1	0.2
Juice	4.8	0.03	1	0.1	0.1	<0.1	0.4	0.3
	9.6	0.06	1	0.3	0.2	<0.1	0.5	0.2
Fruit	4.8	0.03	1	0.6	0.7	0.4	0.3	0.3
	9.6	0.06	1	1.8	1.4	0.6	0.3	0.4
Fruit, young	—	0.06	1	0.5	0.2	<0.1	<0.1	<0.1
	—	0.12	1	1.0	0.3	0.1	<0.1	<0.1
Juice	—	0.06	1	<0.1	<0.1	<0.1	<0.1	<0.1
	—	0.12	1	0.2	0.1	<0.1	<0.1	<0.1
Rind	—	0.06	1	1.7	1.1	0.3	0.1	<0.1
	—	0.12	1	2.7	1.7	0.7	0.3	<0.1
Pulp	—	0.06	1	0.1	<0.1	0.1	<0.1	<0.1
	—	0.12	1	0.2	0.1	0.1	<0.1	<0.1

j) *Pears:* U.S.A. (diethylamine determination)

Type of sample	Conc. (%)	No. appl.	Phosphamidon (p.p.m.)					
			Days after application					
			4	8	12	16	26	31
Fruit	0.04	3	2.8	2.1	1.9	0.6	0.5	0.2

k) *Strawberries:* U.S.A. (cholinesterase inhibition)

Type of sample	Conc. (%)	No. appl.	Phosphamidon (p.p.m.)				
			Days after application				
			0	3–4	5	8–10	20
Fruit	0.12	1	3.2	1.8	0.8	0.7	0.2
	0.06	1	—	0.5	—	0.3	—

Table I (continued)

l) *Tangerines:* U.S.A. (cholinesterase inhibition)

Type of sample	Dosage (kg. a.i./ha.)	Conc. (%)	No. appl.	Phosphamidon (p.p.m.)
				Days after application
				15
Fruit	2.8	0.15	4	< 0.1
	3	0.19	4	< 0.1

m) *Walnuts:* U.S.A. (cholinesterase inhibition)

Type of sample	Dosage (kg. a.i./ha.)	Conc. (%)	No. appl.	Phosphamidon (p.p.m.)		
				Days after application		
				0	7	22
Meat	1.1	0.03	1	—	—	< 0.1
	2.2	0.06	1	< 0.1	< 0.1	< 0.1

n) *Watermelons:* U.S.A. (cholinesterase inhibition)

Type of sample	Dosage (kg. a.i./ha.)	Conc. (%)	No. appl.	Phosphamidon (p.p.m.)
				Days after application
				3
Rind	0.55	0.06	8	< 0.1
Meat	0.55	0.06	8	< 0.1
Rind	0.55	0.06	5	< 0.1
Meat	0.55	0.06	5	< 0.1
Rind	0.55	0.06	3	< 0.1
Meat	0.55	0.06	3	< 0.1

Table II. *Residues of phosphamidon in vegetables and field crops, expressed as p.p.m. after the indicated time intervals*

a) *Alfalfa:* U.S.A. (cholinesterase inhibition or double paper chromatography)

Type of sample	Dosage (kg. a.i./ ha.)	Conc. (%)	No. appl.	Phosphamidon (p.p.m.)						
				Days after application						
				0	1	2–3	4–5	8–9	18	23
Stalks	0.15	—	1	5.1	—	—	0.1	0.1	0.08	0.05
	0.30	—	1	5.5	—	—	0.1	< 0.05	< 0.05	< 0.05
Hay, dry	0.55	—	1	—	—	0.3	0.1	< 0.05	—	—
	—	0.015	3	—	4.3	1.5	—	0.3	—	—

Table II (continued)

b) *Beans:* U.S.A. and England (cholinesterase inhibition)

Type of sample	Dosage (kg. a.i./ ha.)	Conc. (%)	No. appl.	Phosphamidon (p.p.m.) Days after application						
				0	1	2–3	4–6	8–10	12–15	21–25
Pods	0.55	0.06	1	1.8	—	0.6	0.3	—	—	—
	0.55	0.06	5	—	—	—	0.2	0.1	—	—
	0.55	0.06	5	—	—	—	0.23	0.2	—	—
	0.55	0.06	3	—	—	—	<0.1	<0.1	—	—
	0.55	0.06	4	—	—	—	—	—	0.3	0.18
Leaves [a]	0.43	—	1	4.9	1.9	0.9	0.7	0.2	—	—
	0.30	—	1	2.5	1.3	0.8	0.5	0.1	—	—
	0.30	—	1	5.1	1.9	0.65	0.5	0.25	—	—

[a] Ultra-low volume application by aircraft. Samples analysed by *Huntingdon Research Centre,* Huntingdon, England (1968).

c) *Broad beans:* Switzerland (phosphate determination)

Type of sample	Conc. (%)	No. appl.	Phosphamidon (p.p.m.) Days after application					
			1	2	3	6	8	10
Foliage	0.04	1	17.3	9.3	7.4	2.6	1.1	0.1

d) *Broccoli:* U.S.A. (cholinesterase inhibition)

Type of sample	Dosage (kg. a.i./ha.)	Conc. (%)	No. appl.	Phosphamidon (p.p.m.) Days after application			
				0	3	7–9	15
Head	0.55	0.06	—	1.8	0.4	0.3	0.3
	0.55	0.06	4	—	0.32	0.12	—
	0.55	0.06	3	—	<0.1	<0.1	—

e) *Brussels sprouts:* Switzerland (cholinesterase inhibition)

Type of sample	Conc. (%)	No. appl.	Phosphamidon (p.p.m.) Days after application			
			0	5	10	20
Sprouts	0.04	1	0.5	0.13	0.1	<0.1

Table II (continued)

f) *Carrots:* Switzerland (cholinesterase inhibition or phosphate determination)

Type of sample	Conc. (%)	No. appl.	Phosphamidon (p.p.m.) Days after application					
			0	3	6	14	18	28–34
Roots	0.03	1	< 0.1	< 0.1	< 0.1	< 0.1	—	—
	0.025	1	—	—	—	—	—	< 0.1
	0.025	1	—	—	—	—	—	< 0.1
	0.06	3	0.4	0.15	< 0.1	—	—	—
	0.04	1	0.22	< 0.04	—	—	—	—
	0.04	3	—	—	—	—	< 0.05	< 0.05
	0.08	3	—	—	—	—	< 0.05	—

g) *Cauliflower:* U.S.A. (cholinesterase inhibition)

Type of sample	Dosage (kg. a.i./ha.)	Conc. (%)	No. appl.	Phosphamidon (p.p.m.) Days after application	
				3	7
Head	0.55	0.06	4	0.25	< 0.1
	0.55	0.06	3	0.25	0.1

h) *Celeriac:* Switzerland (cholinesterase inhibition)

Type of sample	Conc. (%)	No. appl.	Phosphamidon (p.p.m.) Days after application			
			0	5	10	21
Leaves	0.03	1	—	—	—	0.16
Roots	0.03	1	—	—	—	0.15
	0.03	1	—	—	—	< 0.1
	0.04	1	0.25	0.1	< 0.1	< 0.1

i) *Chard:* Switzerland (double paper chromatography)

Type of sample	Conc. (%)	No. appl.	Phosphamidon (p.p.m.) Days after application				
			1	2	5	8	12
Foliage	0.04	1	27.8	11.3	9.3	1.3	< 0.1

Table II (continued)

j) *Green peppers:* U.S.A. (cholinesterase inhibition or double paper chromatography)

Type of sample	Dosage (kg. a.i./ha.)	Conc. (%)	No. appl.	Phosphamidon (p.p.m.)				
				Days after application				
				0	2–3	6–7	13	20
Fruit	0.55	0.06	1	1.9	0.3	0.1	<0.1	<0.1
	1.1	0.12	1	3.8	0.5	0.35	0.2	0.1
	0.5	—	1	—	0.3	0.1	—	—

k) *Cucumbers:* U.S.A. (cholinesterase inhibition, diethylamine determination, or phosphate determination)

Type of sample	Dosage (kg. a.i./ha.)	Conc. (%)	No. appl.	Phosphamidon (p.p.m.)					
				Days after application					
				0	1	3–4	6–7	14	23
Fruit	0.55	—	1	—	1.2	0.23	<0.1	<0.1	<0.1
	1.1	—	1	—	1.7	0.35	0.27	0.1	0.1
	0.55	0.06	4	—	—	<0.1	0.1	—	—
	0.55	0.06	3	—	—	0.4	0.2	—	—
	0.55	0.06	3	—	—	0.14	<0.1	—	—
	—	0.02	3	0.35	—	<0.1	—	—	—
	—	0.03	3	0.35	—	<0.1	—	—	—
	—	0.06	3	1.25	—	—	—	—	—
	—	0.06	1	1.2	1.6	<0.1	<0.1	—	—
	—	0.03	1	<0.1	—	<0.1	<0.1	<0.1	—
	—	0.03	1	<0.1	—	<0.1	<0.1	<0.1	—

l) *Potatoes:* U.S.A. and Switzerland (cholinesterase inhibition, or double paper chromatography)

Type of sample	Dosage (kg. a.i./ha.)	Conc. (%)	No. appl.	Phosphamidon (p.p.m.)					
				Days after application					
				0	1	5	9–10	12–16	21
Tubers	1.1	0.45	3	—	<0.1	<0.1	<0.1	—	—
	1.1	1.2	3	—	<0.1	<0.1	<0.1	—	—
	—	0.06	1	<0.1	—	<0.1	—	<0.1	—
	0.25	—	7	<0.1	<0.1	—	<0.1	<0.1	—
	0.5	—	7	—	<0.1	—	<0.1	<0.1	—
	—	0.03	1	—	—	—	<0.1	—	<0.1
	—	0.03	—	—	—	<0.1	—	<0.1	<0.1

Table II (continued)

m) *Spinach:* U.S.A. and Switzerland (double paper chromatography)

Type of sample	Dosage (kg. a.i./ha.)	Conc. (%)	No. appl.	Phosphamidon (p.p.m.) Days after application						
				0	1	3	5	8	10	16
Leaves	—	0.06	1	12	5.4	0.4	< 0.1	< 0.1	< 0.1	< 0.1
	—	0.12	1	21	8.1	2.7	< 0.1	< 0.1	< 0.1	< 0.1
	0.25	—	1	12.5	—	—	1.4	0.8	—	0.1
	0.5	—	1	13.2	—	—	2.4	0.5	—	< 0.1
	—	0.02	—	—	30	1.2	—	0.1	—	< 0.1
	—	0.04	—	—	42	0.6	—	0.1	—	< 0.1

n) *Tomatoes:* U.S.A. (cholinesterase inhibition, double paper chromatography, or phosphate determination)

Type of sample	Dosage (kg. a.i./ha.)	Conc. (%)	No. appl.	Phosphamidon (p.p.m.) Days after application						
				0	1	3	4	7	10	15
Fruit	0.55	0.06	3	—	< 0.1	0.1	—	—	—	—
	1.1	0.06	5	—	0.1	0.15	—	—	—	—
	0.55	0.05	4	—	0.3	0.2	—	—	—	—
	—	0.06	1	5.4	3.0	—	< 0.1	< 0.1	—	—
Fruit [a]	—	0.03	3	0.6	—	< 0.1	—	< 0.1	< 0.1	—
	—	0.06	3	0.8	—	< 0.1	—	< 0.1	< 0.1	—
Fruit	—	0.03	3	0.3	—	< 0.1	—	< 0.1	< 0.1	< 0.1
Fruit	—	0.06	3	0.4	—	< 0.1	—	< 0.1	< 0.1	< 0.1

[a] Menzer and Ditman (1963 b).

Table III. *Residues of phosphamidon in cereals, expressed as p.p.m. after the indicated time intervals*

a) *Barley:* U.S.A. (cholinesterase inhibition)

Type of sample	Dosage (kg. a.i./ha.)	Conc. (%)	No. appl.	Phosphamidon (p.p.m.) Days after application
				88
Straw	—	0.4	1	0.16
Grain	—	0.4	1	0.06

Table III (continued)

b) *Rice:* U.S.A. (cholinesterase inhibition)

Type of sample	Dosage (kg. a.i./ha.)	No. appl.	Phosphamidon (p.p.m.)						
			Days after application						
			0	1	3	7	10	14	21.
Foliage	0.4	2	22	10	3.5	0.45	—	< 0.05	< 0.05
	0.4	2	30	17	6.6	1.7	0.6	0.11	< 0.05
Grains	0.55	1	0.9	—	—	< 0.05	—	< 0.05	< 0.05
	1.1	1	1.5	—	—	0.05	—	< 0.05	< 0.05

c) *Wheat:* U.S.A. (cholinesterase inhibition)

Type of sample	Dosage (kg. a.i./ha.)	Conc. (%)	No. appl.	Phosphamidon (p.p.m.)					
				Days after application					
				0	3	10	14	21	60–70
Stalks	0.55	—	1	6.7	1.3	0.16	< 0.05	< 0.05	—
	1.1	—	1	17.5	3.3	0.21	< 0.05	< 0.05	—
Grain	—	0.03	1	—	—	—	—	—	< 0.1
	—	0.03	1	—	—	—	—	—	< 0.1
	—	0.06	1	—	—	—	—	—	< 0.1

Table IV. *Residues of phosphamidon in various other crops, expressed as p.p.m. after the indicated time intervals*

a) *Cotton:* U.S.A. (cholinesterase inhibition)

Type of sample	Dosage (kg. a.i./ha.)	Conc. (%)	No. appl.	Phosphamidon (p.p.m.)							
				Days after application							
				0	1	3	7	10	14	21	28
Foliage	0.3	—	2	78	11	3.5	0.36	0.05	< 0.05	< 0.05	< 0.05

b) *Olives:* Greece and Yugoslavia (double paper chromatography)

Type of sample	Conc. (%)	No. appl.	Phosphamidon (p.p.m.)	
			Days after application	
			8–10	16–20
Fruits	0.03	2	—	< 0.01
	0.04	2	< 0.05	—
Oil	0.03	2	0.06	< 0.01

Table IV (continued)

c) *Sugar beets:* England and Switzerland (cholinesterase inhibition or double paper chromatography)

Type of sample	Dosage (kg. a.i./ha.)	No. appl.	Phosphamidon (p.p.m.)					
			Days after application					
			0	1	2	4	8	16
Foliage [a]	0.43	1	3.2	2.3	0.2	0.1	< 0.1	< 0.1
	0.43	1	5.0	3.5	1.5	0.6	< 0.1	< 0.1
	0.43	1	2.9	2.4	1.2	1.6	< 0.1	< 0.1
Foliage	0.4	3	—	—	—	0.23	< 0.01	—
Beets	0.4	3	—	—	—	< 0.01	< 0.01	—

[a] Ultra-low volume application by aircraft. Samples analysed by *Huntingdon Research Centre,* Huntingdon, England (1968).

d) *Sugar cane:* U.S.A. (cholinesterase inhibition)

Type of sample	Dosage (kg. a.i./ha.)	No. appl.	Phosphamidon (p.p.m.)				
			Days after application				
			0	1	3	7	10
Stalks	0.55	4	0.1	< 0.05	< 0.05	< 0.05	< 0.05

e) *Tea:* Ceylon (double paper chromatography)

Type of sample	Conc. (%)	No. appl.	Phosphamidon (p.p.m.)
			Days after application
			9
Leaves	0.03	2	0.02

f) *Tobacco:* U.S.A. and Mexico (phosphate determination or gas chromatography)

Type of sample	Conc. (%)	No. appl.	Phosphamidon (p.p.m.)			
			Days after application			
			0	2	4–5	18
Leaves [a]	0.03	1	21.6	5.5	< 0.1	—
Leaves (fermented)	0.05–0.08	multiple	—	—	—	< 0.05

[a] Menzer and Ditman (1963 b).

Tables I to IV. Various analytical methods were used for residue determinations, and these are noted in the tables. For further details the reader is referred to the chapter on residue methods.

Since phosphamidon is highly soluble in water, any external residues on fruits and vegetables may be removed by washing. External residues may be expected, particularly on leaves and fruits which are covered with layers of wax. Traces of insecticide within crops are readily degraded by cooking. Residue analyses of apples and pears to which had been added 0.74 p.p.m. phosphamidon and 0.24 p.p.m. desethylphosphamidon showed a breakdown of about 80 percent of the toxic material after one hour of cooking *(California Chemical Company* 1965 a).

The high water solubility of phosphamidon is also responsible for its favourable oil-water partition coefficient, which is important in the treatment of oil crops, such as olives, soybeans, and cotton. Numerous experiments *(CIBA* 1964 a, b) with a wide variety of olives and olive oils showed that the residue level fell to below 0.1 p.p.m. within two or three weeks of the application of the insecticide. Most of the samples analysed contained no detectable residues (< 0.01 p.p.m.). Traces of phosphamidon are further reduced by the processing of the olive oil, which includes treatment with water.

b) Desethylphosphamidon

Residues of desethylphosphamidon in four different California crops were determined after a phosphamidon spray program comprising four applications. The results are summarized in Table V. Again the rate of disappearance was higher in leafy crops than in fruits such as apples and oranges. In the same crop samples, residue levels of the parent compound were always higher than those of the metabolite, except in apples where 0.58 p.p.m. of desethylphosphamidon and 0.33 p.p.m. of phosphamidon were recovered 12 days after the last of four sprayings *(California Chemical Company* 1964 a).

Table V. *Residues of desethylphosphamidon determined in four different crops after phosphamidon application* [a]

Crop	Dosage (kg. a.i./ha.)	Conc. (%)	Desethylphosphamidon (p.p.m.)									
			Days after the last of four applications									
			0	2	4	8	9	12	18	21	25	38
Oranges	—	0.06	—	—	0.33	—	0.23	0.11	—	—	—	—
Apples	3.0	—	—	0.74	0.67	0.64	—	0.58	—	—	—	—
Celery	—	0.06	—	—	—	—	—	—	—	0.57	0.51	< 0.01
Beans, pods	—	0.06	—	—	—	—	—	0.15	—	< 0.01	—	—

[a] Carried out by *California Chemical Company* (1964 a) by cholinesterase inhibition analysis.

c) Other compounds

Desmethylphosphamidon is a metabolite of minor toxicological importance (see chapter on toxicology). Residues of this compound determined as phosphamidon after esterification with diazomethane were extremely small, i. e., 0.45 to 0.61 percent of the initially applied amount of phosphamidon after four days, and 0.04 to 0.15 percent after eight days (*CIBA* 1964 c).

γ-Chlorophosphamidon is a by-product of phosphamidon manufacture, accounting for about two percent or less of the total product. Its mammalian toxicity is much lower than that of the parent compound, and it is a potent inhibitor of cholinesterase. Residues are determined by means of a cholinesterase spray after a suitable cleanup on thin-layer plates. Oranges, apples, celery, beans, wheat, and cabbage were used in these residue experiments. The residues of γ-chlorophosphamidon extracted from the plants were always of the order of magnitude expected, i. e., approximately two percent of the phosphamidon residues. γ-Chlorophosphamidon was degraded at least as rapidly as phosphamidon itself. Phosphamidon did not disappear before γ-chlorophosphamidon (*California Chemical Company* 1964 b).

The two nonphosphorylated metabolites, 2-chloro-N,N-diethyl acetoacetamide and 2-chloro-N-ethylacetoacetamide are of minor importance. When the RM-4C method was used in combination with electron capture GLC to analyse apples, cabbage, and wheat, the residues of both compounds were less than 0.01 p.p.m. throughout the experiment (zero to 56 days). The diethylamide, however, was detected in celery leaves, in which the residues decreased from 0.36 to less than 0.01 p.p.m. within 38 days. The diethylamide was also detected in whole bean plants, in which the residue level decreased from 0.92 to 0.45 p.p.m. within eight days after the last of four applications. The monoethylamide, however, was not detected (*California Chemical Company* 1964 a).

IV. Residues in water, soil, and air

The behavior of phosphamidon in the environment was studied by residue analysis of water, soil, and air. The following data, supplied by the *California Chemical Company* and *CIBA*, demonstrate that from the point of view of degradation rates and residue levels in the environment it can be regarded as relatively safe.

The rates of breakdown of phosphamidon in soils and water were studied in experiments in which known quantities of the insecticide were added to these carriers (Table VI). The half-life of phosphamidon in sea water taken from the San Francisco Bay and kept at ambient temperature was somewhat less than two weeks (*California Chemical Company* 1963). The results from soil experiments varied to some extent, and it seems that large concentrations disappear more rapidly than smaller ones. With loam, loamy sand, and sand the half-lives of phosphamidon were found to be one week or less at a high level of fortification (five p.p.m.), whereas with loam and silt fortified at the one p.p.m. level, half-lives were of the order of three to four weeks (*California Chemical Company* 1965 b). Since field applications of phosphamidon could contaminate the soil and drainage water, its leaching behavior was also investigated. The experiments were carried out with four types of soils fortified with 500 μg. of phosphamidon. The results are presented in Table VII. For phosphamidon residue analysis the soil columns were cut into three sections, each five cm. long, after leaching with 150 mm. of rain. There was no adsorption in samples of sand and sandy loam, and all the added phosphamidon leached through. Likewise with loam the total recovery was good, but the major portion of phosphamidon was found in the bottom segment. With silt, however, total recovery was only 39 percent, the main portion of the compound being recovered from the top zero-to-five cm. layer of the column (*California Chemical Company* 1965 c).

Table VI. *Rate of breakdown of phosphamidon in seawater and different types of soil after single application* [a]

Sample	Dosage (p.p.m.)	Phosphamidon (p.p.m.)									
		Days after treatment									
		0	3	7	14	21	28	35	42	49	56
Seawater	2	—	—	1.6	0.9	0.6	—	0.4	—	< 0.1	—
	5	—	—	3.7	2.1	1.5	—	1.2	—	0.8	—
Loam	1	1.06	1.01	0.95	0.84	—	0.60	—	0.48	—	0.30
	5	4.80	3.68	2.18	1.36	—	—	—	—	—	—
Loamy sand	5	4.76	2.32	1.15	—	—	—	—	—	—	—
Sand	5	4.21	1.17	0.50	—	—	—	—	—	—	—
Silt	1	1.06	1.0	0.84	0.74	—	—	—	—	—	—

[a] Residue values obtained by cholinesterase inhibition analysis (*California Chemical Company* 1965 b).

Table VII. *Leaching of phosphamidon in soil* [a]

Type of soil	Percent of total recovered in fraction				Total percent recovered of amount applied
	Top segment	Middle segment	Bottom segment	Leachate	
Sand	—	—	—	100	94
Loamy sand	—	—	—	100	89
Loam	2	16	75	7	85
Silt	35	27	21	17	39

[a] Analysis of phosphamidon carried out by cholinesterase inhibition (*California Chemical Company* 1965 a).

CIBA (1964 d) studied contamination of the air in apple orchards after spraying with phosphamidon. The first experiment was carried out at high temperature in calm weather. The concentration of active ingredient in the air 18 to 24 hours after spraying was less than 0.005 mg./cu.m., an amount representing no hazard to the health of workers working in the orchards. In a second experiment, done under moderate wind conditions (one to two m./second), no phosphamidon was detectable. Residues on the treated foliage were approximately 20 p.p.m. Similar results were obtained from experiments in a California grapefruit orchard. In this spray experiment air samples were taken at a height of about 1.5 m. above soil level from one to 23 hours after application, the weather being clear, calm, and warm. Phosphamidon residues were always below 0.01 mg./cu.m., at a time when the foliage itself contained more than 150 p.p.m. *(California Chemical Company* 1966).

V. Discussion

The initial half-life of phosphamidon in green metabolizing leaves of various plant species is of the order of one day. Residues in fruits disappear more slowly than those in leaves, but the initial residue level in and/ or on fruits is usually much smaller. In leafy crops, residues generally decrease to a level of less than 0.5 p.p.m. within the first week after normal application, whereas with fruits it may be necessary to wait two to three weeks before this level is reached. Residues in samples of root crops, such as carrots, sugar beets, or potatoes are always below or near the limit of detection. This is not surprising since physiological experiments showed no downward translocation of phosphamidon.

The major metabolite of phosphamidon, *N*-desethylphosphamidon, and the impurity *γ*-chlorophosphamidon, are degraded at the same rate as the parent compound, and so do not accumulate. Residues of the non-toxic metabolite desmethylphosphamidon are extremely small, *i. e.*, less than one percent of the initial phosphamidon residues. Among the non-phosphorylated non-toxic metabolites only 2-chloro-*N*-diethylacetoacetamide was detected in plant material after four applications of phosphamidon. Residues of this compound varied between 0.01 and 0.9 p.p.m., depending on the crop and time of analysis.

Hazardous contamination of the environment does not occur, provided that phosphamidon is applied according to the official recommendations. Half-lives of the compound in soil and in water varied between one and three weeks. Phosphamidon is leached from many soils, so that it does not accumulate in the top soil layers. Samples of air taken in orchards one to 24 hours after spraying contained less than 0.01 mg. of phosphamidon/ cu.m.

Summary

The rates of degradation of phosphamidon, desethylphosphamidon, and *γ*-chlorophosphamidon have been tested under laboratory and greenhouse conditions. All three compounds were found to be rapidly degraded by green metabolising leaves, the half-lives being one to 1.5 days. Fruits, however, degrade phosphamidon at a much slower rate. Residue values, obtained from field experiments during the past years and carried out with a large variety of vegetable, fruit, and cereal crops in many countries of the world, are listed in the present chapter. In leafy crops residues were found to decrease to a level of less than 0.5 p.p.m. within the first week after normal application, but in fruit crops the waiting period can be two to three times longer, although the initial residues are usually small.

Phosphamidon is degraded in seawater, the half-life being approximately two weeks. The rate of degradation in soil depends on the soil type and the initial residue level. The amount of phosphamidon in air collected in orchards after a spray treatment was found to be less than 0.005 mg./cu.m. of air.

Résumé *

Vitesse de dégradation du phosphamidon et importance des résidus

Des essais effectués au laboratoire et en serre, ont montré que la demi-durée de vie du phosphamidon, du deséthylphosphamidon et du γ-chlorophosphamidon, lors de leur application sur les feuilles vertes, est de l'ordre de 1,5 jours. Dans les fruits, la vitesse de dégradation est toutefois plus faible. Ce chapitre rassemble les valeurs obtenues pour les résidus de l'insecticide dans divers légumes, fruits et céréales. Les chiffres ont été tirés d'expériences en plein champ, effectuées au cours des années passées, dans de nombreux pays. Les légumes et les feuilles contiennent moins de 0,5 ppm, une semaine après un traitement normal. Toutefois, pour les fruits, la période d'attente peut être deux ou trois fois plus longue, bien que les concentrations en résidus soient normalement faibles, dès le début. Le phosphamidon est dégradé dans l'eau de mer, la demi-durée de vie étant de l'ordre de deux semaines. Le taux de dégradation dans le sol dépend du type de sol et de la concentration initiale en produit. Des échantillons d'air prélevés dans des vergers, après traitement au pulvérisateur, ne contenaient que 0,005 mg. de phosphamidon par m.³ d'air.

Zusammenfassung **

Die Abbaugeschwindigkeit und Rückstandswerte von Phosphamidon

Phosphamidon, Desäthylphosphamidon und γ-Chlorphosphamidon wurden auf grünen Blättern unter Labor- und Gewächshausbedingungen mit einer Halbwertzeit von ca. 1,5 Tagen abgebaut. Auf Früchten ist die Abbaugeschwindigkeit jedoch viel geringer. Das vorliegende Kapitel hat es sich zur Aufgabe gemacht, Rückstandswerte für verschiedene Gemüse-, Obst- und Getreidearten zusammenzustellen; die Daten stammen aus Feld-

---

* Traduit par les auteurs.
** Übersetzt von den Autoren.

versuchen in vielen Ländern. Im Blattgemüse findet man eine Woche nach der Applikation weniger als 0,5 ppm Phosphamidon, sofern normal appliziert wurde. Die Wartefristen für Früchte können jedoch zwei- bis dreimal länger sein, obwohl die Anfangskonzentrationen normalerweise gering sind.

Phosphamidon wird auch in Meerwasser abgebaut. Hier beträgt die Halbwertzeit ungefähr zwei Wochen. Die Abbaugeschwindigkeit in Böden richtet sich nach dem Bodentyp und der Initialkonzentration. Luftproben aus behandelten Obstanlagen enthielten weniger als 0,005 mg. Phosphamidon/m.³ Luft.

References

Anliker, R., E. Beriger, M. Geiger, and K. Schmid: Über die Synthese von Phosphamidon und seinen Abbau in Pflanzen. Helv. Chim. Acta 44, 1622 (1961).

Brewerton, H. V.: Phosphamidon residues in apples. New Zealand J. Sci. 6, 259 (1963).

California Chemical Co. (now Chevron Chemical Co.), Ortho Division, Richmond, Calif. (U.S.A.): Unpublished report T-504 (1963).

— Unpublished report (1964 a).

— The appearance and decay of gamma-chlorophosphamidon in crops sprayed with phosphamidon. Unpublished report (1964 b).

— Unpublished reports T-828 and T-829 (1965 a).

— Unpublished report T-787 (1965 b).

— Unpublished report T-788 (1965 c).

— Phosphamidon four spray application program (1966).

— Phosphamidon — environmental contamination. Unpublished report (1966).

CIBA Ltd., Agrochemical Division. Basle, Switzerland: Residues of phosphamidon in olives and olive oil. Unpublished report (1964 a).

— Degradation of phosphamidon in plants and its residues in foods and feeds. Unpublished report (1964 b).

— Metabolism of phosphamidon in plants. Identification of desmethylphosphamidon in plant extracts. Unpublished report (1964 c).

— Environment contamination studies in apple orchards after spraying with phosphamidon. Two unpublished reports (1964 d).

— Weitere Untersuchungen zum Verhalten von Dimecron, Carbicron und Nuvacron in und auf Pflanzen. Unpublished report (1968 a).

— Surface stability and penetration of ULV applied Dimecron 100. Unpublished report (1968 b).

Menzer, R. E., and L. P. Ditman: Effect of environmental factors on phosphamidon degradation. J. Agr. Food Chem. 11, 170 (1963 a).

— Phosphamidon residue studies on various crops. J. Econ. Entomol. 56, 88 (1963 b).

Chapter 9

Toxic effects of phosphamidon to insects and mites

By

Volker Dittrich

Contents

I. Introduction

Research on the toxicity of an insecticide to insects and mites can have two major goals. One of them is to develop reproducible criteria of poisonous effects by applying exact methods of dosing and evaluation. The resulting data can serve as a basis for a variety of comparative investigations. The other goal is to predict a poisonous effect under practical conditions, with a high degree of reliability, by exploiting reliable analogies or by exact reproduction of criteria which are decisive for the toxic effect in practice. Both trends in entomological work are interesting, the first for a solid foundation of knowledge on a pesticide, the second for its use. The ways in which to reach both aims differ mainly in the methods of experimentation. Classical toxicological work is based on the dosage- or time-response of test animals to doses of a poison which are measured out exactly and applied to the candidate. Criteria of intoxication are commonly LD_{50} or LT_{50} values. In entomological practice the standardized method of topical application anwers to these specifications. As an equivalent in mite work the slide-dip technique has become the method of choice. It produces an unknown, but constant dosage of the toxicant as a deposit on immobilized individuals. Greater variability than with the above methods must be expected when insects are tested on living plants, because secondary contact through movement on residues or feeding on poisoned parts make the dosage response dependent on individual behaviour patterns of test animals. The further the imitation of field conditions goes, the greater is the variability to be expected. For example, it is very difficult to establish a dosage-mortality relationship between different quantities of granulated insecticide in soil and the response of aphids on a plant rooting in the treated soil sample. Thus, with increasing realism of testing arrangements, the judgement of results becomes a problem requiring statistical interpretation.

In this work the main attention will be given to results obtained with phosphamidon and comparative compounds which were a direct result of a toxic dose and which were produced with exact laboratory methods. Some typical tests with a more practical approach will also be discussed. In most of the work referred to specificity of toxic effects is what interests us, *i. e.*, that of a new preparation as compared to a standard, that of one pest species to another, of specific oral toxicity compared to contact toxicity. Unchanged toxicity against resistant forms as compared to sensitive ones is a highly attractive feature, and in reversal to above-quoted ends, so is relative non-toxicity to beneficial insects *versus* toxicity to pests. All these juxtapositions point to a main trend in toxicology work: to uncover selective action, to define and use it. This work is an attempt to demonstrate the principle.

II. Results

a) *Toxicity against various pests*

The definition of the relative toxicity of a substance to one or several test species is an important aim of practical research. The simplest way of doing it is to compare the percentage kill of a fixed dosage or concentration with that produced by a standard. This is the usual procedure for the screening of newly synthesized chemicals applied by the manufacturers of pesticides. A more precise variant is to compare LD_{50}-values of two materials after Ldp-relationships have been established. SUN (1950) pointed out that the relative toxicity between standard and candidate is practically constant and not subject to changes in testing techniques or vigor of test animals. Thus, the toxicity index LC_{50} standard/LC_{50} candidate = 100 is a valuable tool to define the toxicity of two insecticides tested at the same time with the same method. If toxicity relationships have been established for several pest species, important conclusions can be made as to the selective action of a compound or a group of related compounds to the pest.

From Table I some important facts can be read. The Noctuid *Barathra brassicae* shows comparatively little sensitivity to the compounds tested which are obviously more toxic for *Musca domestica* and *Chilo suppressalis*. Phosphamidon has a selective toxicity for larvae of the Pyralid moth (*C. suppressalis*) and at the same time the lowest activity against the Noc-

Table I. *Toxicity of phosphamidon and comparative compounds to the housefly, cabbage armyworm, and the striped rice stemborer* (after KOJIMA 1961)

Insecticide	LD$_{50}$ values (µg./g. of insect) [a]		
	♂ *Musca domestica*	*Barathra brassicae* (l.) [b]	*Chilo suppressalis* (l.) [b]
Phosphamidon	19.5	> 500.0	2.0
Phosdrin	1.0	12.4	1.2
DDVP	0.8	42.2	10.4
Demeton methyl	43.5	—	19.5
Methyl parathion	1.3	28.5	0.5
Parathion	1.0	15.6	3.5
Diazinon	1.9	—	2.4

[a] Dosage applied in one µl. of acetone.
[b] l. = larva.

tuid moth *(B. brassicae)*. An interpretation of this phenomenon is difficult. It probably rests on the fact that penetration into the body and detoxification at the site of action vary from one species to the other. The final toxic effect is dependent on the available dosage at the sensitive site which produces the biochemical lesion (SUN 1969). Another important selectivity in Table I is that of DDVP on the Diptera *Musca domestica,* while this material is of lesser toxicity to the lepidopterous species.

From KOJIMA's and NAGAE's (1958) work it is apparent that a direct correlation between the mortality of stemborer larvae which have recently bored into the stem and dosage of toxicant can be established, even though the spray is applied to the plant's surface. They allowed egg batches to hatch on the test plants and treated the plants and mining larvae five days after infestation (Table II).

A further example of an exact definition of toxicity on plants is the establishment of a time-mortality relationship between phosphamidon and

Table II. *Effect of phosphamidon and methyl parathion on mining larvae of the rice-stemborer Chilo suppressalis* (after KOJIMA and NAGAE 1958)

Insecticide	Conc. (p.p.m.)	No. larvae tested	Mortality (%)	LC$_{50}$ (p.p.m.)
Phosphamidon	200	163	84.0	
	100	124	73.4	
	40	145	42.8	55.0
	20	151	15.9	
	10	101	5.9	
Methyl parathion	200	127	92.1	
	100	150	75.5	
	40	145	87.6	19.0
	20	172	52.3	
	10	113	36.3	

malathion and adult ♀♀ of the green rice leafhopper *(Nephotettix bipunctatus)*, also by Kojima and Nagae (1958). In the control of this insect a quick knock-down is essential as it is a vector of viruses. Therefore time-mortalities are preferred to dosage-mortality relationships and the medium lethal time (LT_{50}) of a given dosage is used as the criterium of toxicity. Twenty test specimens were placed on potted rice plants and sprayed under a cage of wire mesh. Every ten minutes paralyzed animals were registered. The trial was replicated four times and the mean percentage kills calculated in percent (Table III).

Table III. *Mean Kd-values of ♀ green rice leafhoppers Nephotettix bipunctatus after treatment with 250 p.p.m. of phosphamidon and malathion* (after Kojima and Nagae 1958)

Insecticide	Conc. (p.p.m.)	Kd_{50} (min.) [a]
Phosphamidon	250	43.0
Malathion	250	21.0 26.6

[a] Combined values.

    In both trials phosphamidon required a longer knock-down time against the leafhopper than malathion which was the standard prior to the onset of resistance against organophosphorus compounds.

    Both experiments of Kojima and Nagae referred to above had a distinct relationship to practical conditions through inclusion of plants into the testing procedure. Their definition of toxicity, however, was made with the classic tools of toxicology.

    Hough's (1962) experimental design is in vivid contrast with this principle. He tested phosphamidon and comparative products in a procedure as close to practical conditions as possible and with the simplest of procedures to evaluate toxicity. His aim was to find whether sprays of standard insecticides at recommended dosages would kill young larvae of the codling moth *Carpocapsa pomonella* which had recently entered immature apples. Egg masses were used for the infestation and those larvae hatching 12 to 14 hours after initiation of the trial. Late hatches were eliminated. The apples were treated in batches of 20 under field conditions and dissected eight to 12 days later for control of mortality. The result is recorded as percent killed of the total number of larvae entered in the fruit (Table IV). These data have considerable significance for the particular problem they were meant to investigate. The conditions of infestation, application of the toxicant, and the differing concentrations used for the sprays were all designed to match practical conditions. On the other hand, such procedure implies that basic results on the comparative toxicity of the test materials cannot be gained under the given conditions.

Table IV. *Effect of phosphamidon and other insecticides on young larvae of the codling moth, Carpocapsa pomonella, mining in immature apples* (abbreviated after HOUGH 1962)

Insecticide	Control after days (in % mortality)						
	1	2	3	4	5	6	Check
Phosphamidon, 580 p.p.m	97.5	—	85.5	—	77.8	—	0.0
Parathion, 360 p.p.m.	97.0	96.2	85.0	79.5	67.6	18.4	0.0
Malathion, 750 p.p.m.	98.4	98.4	87.0	77.4	55.8	19.1	0.8
Carbaryl, 1,200 p.p.m.	97.0	71.5	61.2	40.3	24.2	1.6	1.4

However, an unbridgable gap does not exist between "classical" and "applied" methods of defining toxicity. There are possibilities for making realistic predictions of the toxic effects of pesticides in the field using results obtained with laboratory techniques when an intermediate definition on plants is made using the same criteria of evaluation as with the laboratory methods; thus, in the tests to follow, ldp-lines for phosphamidon and comparative compounds were established for two mite species, *Tetranychus urticae* and *Tetranychus telarius*. We used two methods, the "slide-dip-" and the "leaf-dip-method" (DITTRICH 1962) and extrapolated LC_{50} values. For each insecticide tested the factor LC_{50} SD/LD reflects the influence of the plant on the test result, as the test animals move over treated surfaces or ingest poisoned plant juices in the leaf-dip-method. In the slide-dip-method, it is exclusively the primary contact of the toxicant

Table V. *Influence of test plant on LC_{50} in p.p.m. of phosphamidon and comparative compounds tested against an OP-sensitive Tetranychus urticae and a tolerant T. telarius*

Acaricide	LC_{50} (p.p.m.)					
	Tetranychus urticae			*Tetranychus telarius*		
	LC_{50} SD [a]	LC_{50} LD [b]	$\dfrac{LC_{50} \text{ SD}}{LD}$	LC_{50} SD [a]	LC_{50} LD [b]	$\dfrac{LC_{50} \text{ SD}}{LD}$
Phosphamidon	30.00	6.00	5.0	1,500.00	660.00	2.3
Dicrotophos	12.70	3.20	4.0	1,080.00	490.00	2.3
Amidithion	21.30	14.00	1.5	273.00	52.80	5.2
Oxydemeton methyl	4.75	0.75	6.3	490.00	133.00	3.7
Chlorphenamidine	430.00	340.00	1.3	570.00	410.00	1.4

[a] SD = slide dip.
[b] LD = leaf dip.

with the mite's surface which is responsible for the intoxication of the test animal (Table V).

Similar tests were also made by ELDEFRAWI et al. (1965) who established a relationship between results obtained with the slide-dip versus a leaf-spray method. If extensive knowledge on toxicity response of different mite populations in the field is available on the basis of slide-dip results, a valid recommendation for the use of acaricides can be given by considering the "plant-factor" SD/LD. This example is not a theoretical one. GHOBRIAL et al. (1969) have analyzed the resistance status of mites in Egyptian cotton and acquired an extensive amount of toxicological data in slide-dip results. Such data might also be valuable for greater economy in practical control measures.

b) Toxicity resulting from various modes of application

The last section dealt mainly with the selective toxicity of one chemical to different pest species. We are now going to discuss selective toxicity to one species which is due to a different route of entry into the animal's body. The subject has received necessary attention already in the past, and a concise summary of the complex processes involved in this type of selectivity is that of SUN (1969) who defines it as based on penetration and decomposition. WIESMANN (1960) suggested specific oral toxicity be used in counteracting the development of resistance in houseflies. He pointed

Table VI. Comparison of contact and oral toxicity of phosphamidon and comparative compounds against ♀ houseflies

Insecticide	LD_{50} values (μg./g. of fly)			
	LD_{50} oral	Range LD_{50} oral	LD_{50} topical	Range LD_{50} topical
Phosphamidon	6.19	4.54– 8.26	8.67	7.43–9.50
Dicrotophos	6.19	4.95– 7.43	9.09	8.26–9.91
Dimetilan	59.91	43.38–81.81	7.02	6.19–7.43
Dioxacarb	24.79	19.83–30.16	6.19	5.78–7.02

out that contact with residual deposits of various levels of toxicity was the cause of selection of resistant strains, while active incorporation of the toxicant in a bait would lead to concentrations of the toxicant that excluded survival of the flies. Such a principle had formerly been used with the arsenicals, and its basis was high oral and negligible contact toxicity. With Musca domestica we tried to find such selectivity for several insecticides. Three-to-five-day-old female flies were narcotised under CO_2 and received the toxicant in one μl. of acetone on the dorsum. For the oral tests, the dorsal parts of narcotised specimens were dipped in liquid paraffin of low melting point and the flies were then fixed in groups of ten on microscopic slides. For about ten hours afterward, the animals did not receive food or liquid, but had sufficient humidity in the covering petri dish. The hungry flies then were fed the toxicant dosage in one µl. of a five percent sugar solution by means of a specical syringe with a total volume of one µl. The evaluation was made 24 hours later (Table VI).

In this test the enolphosphates show a different toxicity pattern from that of the carbamates. Their oral toxicity is greater than the contact toxicity while the carbamates show the opposite. Dimetilan, in particular, is 8.5 times more effective in the contact test than in the oral test. This difference might be due to a greater sensitivity of carbamates to the aggressive digesting enzymes of the fly-intestine. Variability of results is about twice as high in oral tests as in topicals. This might be explained by regurgitation of stomach contents which was more frequently observed with carbamates than with phosphates. It does not seem conjectural to attribute this phenomenon to the fly's sense of taste, even though this would cast some doubt on the validity of the principle of bait poisoning in housefly control. Our results on dimetilan having an 8.5 times stronger contact than stomach effect are in contrast to those of WIESMANN (1960), who claimed the oral efficacy to be four times greater than the topical.

Selective oral toxicity can also be important in other insect groups and the ability to avoid intoxicated food by the sense of taste is of great practical consequence. PARRY and FORD (1969) could show that the aphid *Myzus persicae* is able to recognize phosphamidon in sugar solutions and to avoid it. The authors used MITTLER's and DADD's (1964) choice chambers and fed the test animals through Parafilm-membranes. In this arrangement

Table VII. *Avoidance reaction of the aphid Myzus persicae to phosphamidon-poisoned food* (after PARRY and FORD 1969)

Chamber A [a]	Chamber B [a]	No. of salivary plugs
10% sucrose	—	33.4 ± 7.93
—	10% sucrose	32.0 ± 6.75
10% sucrose + 6 p.p.m. P	—	55.6 ± 21.47
—	10% sucrose + 6 p.p.m. P	58.8 ± 14.79
10% sucrose	—	39.8 ± 9.43
—	10% sucrose + 6 p.p.m. P	78.0 ± 17.38
10% sucrose + 9.5 p.p.m. P	—	46.8 ± 2.98
—	10% sucrose + 9.5 p.p.m. P	41.8 ± 5.08
10% sucrose	—	43.0 ± 10.12
—	10% sucrose + 9.5 p.p.m. P	92.0 ± 12.52
10% sucrose + 15 p.p.m. P	—	44.4 ± 16.20
—	10% sucrose + 15 p.p.m. P	55.8 ± 24.77
10% sucrose	—	44.0 ± 14.90
—	10% sucrose + 15 p.p.m. P	82.8 ± 13.58

[a] P = phosphamidon.

the aphids could not identify the poison except by probing with their stylets through the membrane. With each insertion the animals leave traces of saliva which were counted and used as criteria for the frequency of probing. The figures in Table VII were calculated on the basis of five replicates with 20 aphids each.

Table VII gives conclusive evidence that *Myzus persicae* is able to discriminate between sugar solution alone and with phosphamidon added. Since aphids and other highly specialized plant-sucking pests are mainly controlled by systemics, these findings have practical implications. The most dangerous feature of homopterous insects is their potential to transmit plant viruses, and an increased number of probes on a plant means an increasing hazard of virus infection with some viruses. It is conceivable, therefore, that the future development of specific aphicides will have to make allowance for this fact by testing the avoidance reaction of aphids for an additional new selectivity: non-recognizability by taste.

c) Toxicity against beneficial insects

The last paragraph dealt mainly with selective toxicity against pest insects. In recent years, however, a new trend in research on selective toxicity has developed. Ecology has begun to play an important role in plant protection and accordingly the investigation of selective toxicity on beneficial insects has begun to receive the necessary attention. The aim of research here is the opposite of that discussed in the last section. While there it was to choose selectively toxic substances from others, here it is to choose selectively non-toxic compounds from the range of established insecticides. For obvious reasons, it must be a choice among highly active insecticides because of their good toxic properties against insects. With such a paradoxical aim, research is difficult. The chances of finding selective nontoxicity are more than scanty owing to the fact that there are many species classified as "parasites and predators" with a variety of adaptations to their environment. On the other hand, there is the common biochemical mode of action of modern insecticides to which the predators answer as well as the pest. If systemic preparations are dealt with and they are used according to the definition of a systemic material, there is a theoretical possibility of sparing beneficial insects; however, so far in plant protection, the practice of applying systemics to the basal parts of the plant in order to control sucking and biting pests has remained exceptional. The differential behaviour of residues on plant surfaces, which Voss (1968) showed for the enolphosphates phosphamidon, dicrotophos, and monocrotophos, may also have in consequence some selectivity for certain insects which only live on plants, rather than feed on them. But, all in all, it must be expected that a good insecticide will also live up to its reputation against beneficial insects, and results in the literature confirm this. LINDGREN and RIDGWAY (1967) tested the contact toxicity of different preparations on various predators and parasites and their immature stages. They found certain differences in toxicity between compounds and from one insect to another; however, it must be doubted if this is of any practical importance (Table VIII).

The highest toxicities were registered in this test series for dicrotophos and methyl parathion. Phosphamidon took the middle position and demeton was the least toxic material. As the maximum dosage to kill an individual varies considerably owing to the diverging individual weights, the above values are misleading when used to judge ecological effects of insecticide treatments. Logically, LINDGREN and RIDGWAY (1967) have not quoted their LD_{50} values in µg./g. of insect but in µg./insect. The above table was transposed to fit better into the frame of the present article.

Table VIII. *Toxicity of phosphamidon and comparative compounds to various predators. LD_{50} in µg./g. insect. Topical application in one µl. acetone* (after LINDGREN and RIDG-WAY 1967)

Predator	LD_{50} (µg./g. of insect) [a]			
	Phos-phamidon	Dicrotophos	Methyl parathion	Demeton
Chrysopa spec., lacewing (adult)	4.80	2.65	1.08	13.22
Chrysopa carnea, lacewing (larva)	56.40	66.00	290.00	290.00
Collops balteatus, soldier beetle (adult)	106.00	30.60	0.93	147.00
Geocoris puncticeps, lygeid bug (adult)	17.80	5.44	2.60	72.90
Hippodamia convergens, lady beetle (adult)	7.15	0.95	1.35	22.00
Hippodamia convergens, lady beetle (larva)	12.20	2.20	3.17	35.00
Nabis americoforus, bug (adult)	23.60	5.00	2.90	147.30

[a] Topical application in one µl. of acetone.

A series of investigations by BARTLETT (1963, 1964 a and b) was concerned with the effect of insecticide residues on treated surfaces on beneficial insects. He used waxed paper for models of citrus leaves and worked with a standardized residue technique. Twenty-four hours after treatment, the test animals were caged on the residues and kept there for one to three days. In order to evaluate toxic effects he used the symbols H = heavy, M = medium, L = light, O = no effect (Table IX).

From these results it seems more or less impossible to class any of the candidates as particularly harmless to the predators; but the behaviour

Table IX. *Residual toxicity of phosphamidon and comparative preparations to various parasites and predators* (compiled from BARTLETT 1963, 1964 a and b)

Insecticide	Residue (μg./cm.²)	5 Hymenopterous species [a]	6 Coleopterous species [a]	1 Phytoseid mite [a]	1 Neuropterous species [a, b]		
					Egg	Larvae	Adult
Phosphamidon	3.22	5 H	3 M, 3 H	H	O	H	H
Dimethoate	6.44	5 H	1 L–M, 2 M, 3 H	H	O	L	H
Parathion	6.44	5 H	1 M–H, 5 H	H	O–L	H	H
Diazinon	6.44	1 M, 4 H	1 L–M, 1 M, 1 M–H 3 H	H	O	H	H

[a] H = heavy, M = medium, L = light, O = no effect.
[b] A lacewing.

of the residues on plants must not be disregarded in this context. True, the mobile forms are almost completely wiped out, but if, on the other hand, quiescent stages or eggs are not affected by the holocaust, a rather quick reestablishment of the populations may be possible, if toxic residues disappear quickly. According to these views, parathion with its good ovicidal effect and comparatively stable residues looks unfavorable. Dicrotophos and phosphamidon with surface residues which disappear comparatively rapidly (Voss 1968) should be better preparations in the ecological sense.

Of particular importance is the toxic effect of insecticides on honey bees, the chief pollinators of many important cultures. Not only are external residues of concern, but also systemic ones, as these are secreted into the nectar and thus passed on to the foraging bees (JAYCOX 1964). The acute oral toxicity of a great number of current insecticides was investigated by BERAN (1965) and related to the contact toxicity after tarsal exposure to standardized residues (Table X). All four preparations are categorized by BERAN as toxic to bees, but for phosphamidon the danger should be mainly restricted to the ingestion of intoxicated nectar and to a much lesser degree of residues on the plant surface.

Table X. *Oral and contact toxicity of phosphamidon and comparative compounds to the honey bee, Apis mellifica* (after BERAN 1965)

Insecticide	LD$_{50}$ (μg./g. oral)	LD$_{50}$ (μg./100 cm.² tarsal)	Tarsal/oral
Phosphamidon	1.72	148.10	86.0
Phosdrin	0.24	0.98	4.1
Dimethoate	0.79	145.80	168.0
Diazinon	0.84	9.56	11.4

d) Toxicity against normal and resistant races

In a work on the toxicity of phosphamidon and its chemical relatives, a short discussion on the behaviour of these compounds against resistant races of insects and mites has to be included. According to BROWN (1967) 225 resistant insect or acarina species have already been reported; 54 of those are resistant to organophosphorus compounds (OP). The reservoir of resistant species is continuously enlarged by incessant treatments, and new insecticidal groups which act on other than the current biochemical mechanisms are rare.

Under these circumstances, every single compound promising relief is valuable, even if the effect is only temporary. There is no choice but to use compounds the chemical relatives of which may already be embraced by a "group resistance" in some insect species. If this does not, in the long run, give any justifiable hope for solving the resistance problem, it may give instead some more time to search for new active groups against which resistance does not yet exist.

To search for such compounds, the principle of relative toxicity, this time between sensitive and resistant strains of test animals, was applied. According to the LC_{50}-values in both forms, a resistance-index can be calculated which corresponds to SUN's (1960) cotoxicity-coefficient:

$$LC_{50}\ R\text{-strain}/N\text{-strain} = \text{Resistance Index}$$

The trials described below (Table XI) were made with an OP-sensitive strain of the spider mite, *Tetranychus urticae*, and an OP-tolerant *T. telarius* (carmine mite) using the slide-dip method.

The resistance-index LC_{50} T.t./T.u. shows that the protective mechanism in *Tetranychus telarius* is effective against the organophosphates without exception; however, within the group of structurally closely related enolphosphates, there are distinct differences of reaction: the index-value

Table XI. *Toxicity of phosphamidon and comparative preparations against sensitive and OP-resistant spider mites*

Acaricide	LC_{50} values (p.p.m.)		
	Tetranychus urticae sensitive	*Tetranychus telarius* tolerant	*T. t./T. u.* resistance-index
Phosphamidon	30.0	1,500.0	50.0
Dicrotophos	12.7	1,080.0	85.0
Monocrotophos	2.4	64.4	26.6
Amidithion	21.3	273.0	12.9
Parathion	124.0	2,400.0	19.3
Paraoxon	270.0	2,200.0	8.2
Chlorphenamidine	430.0	570.0	1.3

shows falling tendency in the sequence dicrotophos > phosphamidon > monocrotophos. The latter is the most toxic compound against both sensitive and tolerant species. Experience in the field shows that on-plant-treatments have similar results.

Parathion and paraoxon differ visibly in their effect from the enolphosphates. This is obvious from much higher LC_{50}-values in both species. Chlorphenamidine toxicologically belongs to neither group; its resistance index is practically one. As for all other preparations 24-hour values were given for chlorphenamidine which is unfavorable because of its slow mode of action. An intoxication of spider mites with OP-compounds reaches stable values after about 16 hours; for others, like dicofol and chlorphenamidine, it takes about 48 hours until such a stage is reached. In a dosage-mortality relationship this is expressed in lower LC_{50}-values if after treatment holding is extended from 24 to 48 hours. With chlorphenamidine, both species thus show an LC_{50} one decimal lower after 48 than after 24 hours.

Table XII. *Toxicity of phosphamidon and comparative preparations against sensitive and resistant strains of Boophilus microplus, using larval immersion technique* (after SHAW *et al.* 1968)

Insecticide	LC_{50} values (p.p.m.)		
	LC_{50} strain Z (sensitive)	LC_{50} strain L (resistant)	LC_{50} L/Z (R-index)
Phosphamidon	670.0	4,700.0	7.00
Dicrotophos	41.0	1,500.0	37.00
Parathion	0.5	960.0	1,920.00
Formamidine	1.2	0.9	0.75

A similar study of the resistance situation in strains of the cattletick, *Boophilus microplus*, was made by SHAW *et al.* (1968) (Table XII). Both R-indices of the enolphosphates agree with those in our mite species as far as the sequence is concerned. The LC_{50}-values within both strains also have a similar tendency, *i. e.*, phosphamidon > dicrotophos. Parathion, on the other hand, seems to have a stronger effect on tick larvae than on mites, since the figures for Z- as well as for L-animals are relatively low, but also in ticks the values for chlorphenamidine point to other than the OP-mechanism as a cause for this material's toxicity.

To sum up: it is apparent that all OP-compounds tested here were subject to the protective mechanism devised by the resistant strains used. However, the results demonstrate that within the enolphosphates containing a substituted amino-group, the protective mechanism does not function equally well. Monocrotophos is the most active substance against resistant forms.

e) Toxicity due to mixtures with other chemicals

The most important toxicological interactions to be treated in this chapter are synergism and antagonism. According to METCALF (1967), synergism is a joint toxicological or pharmacological effect which is substantially greater than the additive action of the components. Accordingly, antagonism is a lesser effect from a mixture than the more active component would produce alone. There is general agreement that synergists are compounds which are ineffective when used alone at the same dosages as when effectively applied jointly with a synergand. Synergists might have the following remarkable effects: 1) decrease the costs of expensive insecticides, 2) widen the spectrum of activity, and 3) restitute activity lost through resistance.

Together with that of other enolphosphates, the synergisability of phosphamidon was investigated quite some time ago. SUN and JOHNSON (1960) in their tests with methylenedioxy-compounds of the type piperonyl

Table XIII. *Synergism by combinations of OP-compounds with sesamex in tests with Musca domestica* (after SUN and JOHNSON 1960)

| OP compound | LC$_{50}$ values (%) | | | Cotoxicity coefficient with 1% sesamex |
	Toxicant	Toxicant + 1% sesamex	Ratio sesamex/ toxicant	
Phosphamidon	0.024	0.00165	600	14.6
Dicrotophos	0.059	0.003	333	19.7
Methyl parathion	0.0053	0.013	72	0.41
Methyl paraoxon	0.0039	0.002	455	1.8

butoxide were able to demonstrate that there are typical interactions between the above-mentioned classes of chemicals. For their trials with the housefly, they used a spraying technique with kerosene as solvent, the flies being treated in cages. The concentration of their synergist, sesamex, was kept at a constant one percent, independent of the concentration of active ingredients used. From the Kd-values, Kd$_{50}$ values were plotted and used as an expression of synergism in a particular form: Kd$_{50}$ toxicant alone/Kd$_{50}$ toxicant in mixture = cotoxicity coefficient. According to Table XIII organophosphorus compounds of differing structures show a varying degree of synergism when combined with the same synergist. High cotoxicity-coefficients for the enolphosphates indicate strong synergism and the negative figure for methyl parathion antagonism. Tests with other insect and mite species produced corresponding results; the intensity of reaction to the sesamex-fortified preparations, however, varied from one species to another. Generally, the sensitivity to synergized enolphosphates was:

housefly ≫ spider mites ≧ aphids.

If the ratio sesamex/toxicant is compared on the basis of the LC$_{50}$ in the mixture, it is apparent that this is an extremely volatile figure from one treatment to next. This results from the fact that the authors, for the sake of greater experimental simplicity, always kept the sesamex-concentration in their tests at one percent; thus, a fixed insecticide/synergist ratio was neither achieved nor intended. The consequence of varying ratios was absolutely clear to the authors, as their own experiments show increasing cotoxicity-coefficients when insecticide concentrations are fixed but the sesamex percentage increased.

ZSCHINTZSCH's (1961) results on synergism between methylenedioxy compounds and phosphamidon agree with those above-mentioned; however, he used a different technique to demonstrate synergism. This consisted of exposing *Drosophila melanogaster* to residues of synergist and insecticide and establishing time-mortality relationships. It is also clear from ZSCHINTZSCH's results that the degree of synergism depends on the two partners. Combinations with fixed concentrations of insecticide and increasing concentrations of synergist resulted in higher toxicity and *vice versa*. A new point in these investigations was the demonstration of a shift from synergism to antagonism as a consequence of continuously varying the concentrations of both components in the mixture.

ZSCHINTZSCH classified several types of insecticides which show typical interactions with methylenedioxy compounds:

1) synergism over the whole range of concentrations;
2) antagonism over the whole range of concentrations; and
3) antagonism in high, synergism in low concentrations.

The following Table XIV is a summary of a great variety of tests and presents the data with simplifying symbols.

Table XIV. *Interaction of methylenedioxy compounds with OP-insecticides from time-mortality data* (after ZSCHINTZSCH 1961) [a]

Insecticide	Piperonyl butoxide	Sulfoxide	Piperonyl cyclonene	Propylisome
Phosphamidon	S	SS	S	S
Phosdrin	SS	S	SS	SS
Paraoxon	SS	SS	SS	SS
Chlorthion	AAA	AAA	AAA	AAA
Thiometon	AA	AA	AA	AA
Parathion	AA → SS	AA → SS	A → SS	A → SS
Diazinon	A → SS	A → SS	A → S	A → S

[a] A = antagonism, S = synergism, A → S = antagonism in high concentrations of toxicant, synergism in low.

In Table XIV phosphamidon belongs to those compounds which are synergised by all synergists at all concentration levels. With chlorthion and thiometon the opposite is true; but toxicologically most remarkable is the behaviour of parathion and diazinon. Both shift between the extremes of synergism and antagonism depending on the concentrations of the partners which are both varied in these trials. According to SUN and JOHNSON (1960) sesamex blocks the activation $P = S \rightarrow P = O$ in thionophosphates by inhibition of the microsomal oxidases and at the same time stabilises the $P = O$ metabolite.

An interpretation of a mechanism of synergism with the vinylphosphates was also tried by these authors but less successfully. They speculated that by desalkylation on the nitrogen less toxic metabolites were formed and synergism originated when such a metabolic step was inhibited by the presence of a synergist. Unfortunately, such an elegant solution is not at hand, as HALL and SUN (1965) demonstrated implicitly in their experiments with dicrotophos and its desalkylation product monocrotophos. Our own results also show that the desalkylation products are more toxic than the parent compounds (Table XV).

Table XV. *Toxicity of phosphamidon and dicrotophos and their N-desalkylation products on female houseflies*

Compound	LD_{50} (µg./fly)	LD_{50} (µg./g.)	Range (µg./g. 95 c.i.)
Phosphamidon	0.210	8.7	7.5– 9.5
N-Ethylphosphamidon	0.160	6.6	5.8– 7.1
Dicrotophos	0.220	9.2	8.3–10.0
N-Methyldicrotophos	0.066	2.7	2.4– 3.1

ZSCHINTZSCH's (1961) observations of a startling reversal in toxicological behaviour of synergistic mixtures are not an isolated case. Cooresponding results were also reported by MÜLLER (1967) with an N-methyl carbamate and piperonyl butoxide. These incidences show that care has to be taken when the usual 1 : 5 insecticide-synergist ratios are used for experimentation as is usual since MOORFIELD's (1958) work. Cotoxicity-coefficients are dangerous when used to characterise synergism because they do not tell the complete story. Also, for practical purposes, a better definition of synergism is necessary in case mixtures should ever be produced and used on a large scale. Only if such a method existed could preparations be produced which were exactly defined in their activity and which would be balanced accordingly.

An exact definition of a synergistic relationship can be made by establishing an isobole, a curve of equal toxicological effects. It is constructed by testing a multitude of synergistic mixtures with a constant dosage of one and a variable number of dosages of the other partner. Both the constant and the variable dosages follow logarithmic progression. From the resulting ldp-relationship LD_{50}-values are extrapolated which form points of the isobole. If there is synergistic action the isobole will have a parabolic shape and its apex will be turned toward the starting point of the coordinate system. This presentation of toxicological or pharmacological interaction was first described by Loewe and Muischnek (1926). Tammes (1964) demonstrated its merits for agricultural trials.

The isoboles of phosphamidon and dicrotophos in comparison to dioxacarb were defined using C-19795 as the synergist. C-19795 is 2,6-dichlorobenzaldoxime propynyl ether and is covered by patent for *CIBA*. It contains an acetylene group which was described as an important grouping for synergistic action in propynyl ethers by Kooy (1966).

Naphthyl compounds containing this group were investigated as synergists for carbaryl by Sacher *et al.* (1968). For our trials three-to-five-day-old female houseflies were used. They had been kept on a diet of sugar prior to testing. They received their dosage under CO_2 narcosis in one µl. of acetone in groups of ten. For the following 24-hour holding period the flies received honey water on a plug of cotton wool for food and liquid supply. Each mean in the following Table XVI is composed of Kd-values for $4 \times 2 \times 10 = 80$ treated flies, with treatments distributed over four days.

Table XVI. *Mean percentages knock-down of ♀ houseflies after treatments with mixtures of phosphamidon and C-19795*

C-19795 (µg./fly)	Knock-down (%)						
	Phosphamidon (µg./fly)						
	0.05	0.07	0.10	0.14	0.20	0.27	0.37
1.60	95	—	—	—	—	—	—
0.80	80	100	—	—	—	—	—
0.40	72	86	93	—	—	—	—
0.20	53	82	70	97	—	—	—
0.10	25	50	50	70	88	—	—
0.05	5	20	30	40	66	87	—
0.025	—	—	10	23	36	72	—
0.012	—	—	—	16	43	60	—
0.006	—	—	—	—	33	62	—
0.003	—	—	—	—	—	62	—
0.0015	—	—	—	—	—	59	—
0.00	—	—	13	20	43	60	87

From the Kd-percentages in this table ldp-lines were drawn, the columns representing a constant phosphamidon dosage and variable C-19795 dosages. In the rows there was a constant C-19795 dosage and variable phosphamidon dosages. Ldp-lines were graphically determined and resulted in LD_{50}-values as presented in Fig. 1. For dicrotophos and dioxacarb an identical procedure was used.

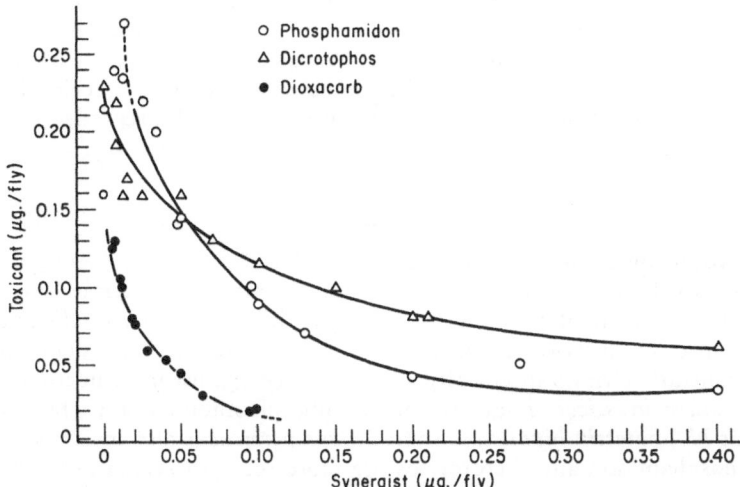

Fig. 1. Isoboles demonstrating synergism between C-19795 and the insecticides phosphamidon, dicrotophos, and dioxacarb

The shape of the three isoboles indicates essential differences in synergisability of the three compounds by the same synergist. The curve of the carbamate is almost perfectly symmetrical, thus allowing substitution of each partner by the other to an almost unlimited extent. Since the toxicity of the unsynergised carbamate is greater than that of the enolphosphates, the isobole of the carbamate comes closest to the zero-point of the coordinate system.

The curves of both enolphosphates are quite distinct from that of the carbamate. In its shape the phosphamidon curve has a certain affinity with that of dioxacarb, *i. e.*, it is a comparatively symmetrical parabel. In the high synergist-low toxicant range, phosphamidon is not so effectively synergised as dioxacarb, but significantly better than dicrotophos. There is a similar situation in the high toxicant-low synergist dosages. Even tiny synergist dosages will cause definite increases in toxicity reaching byond the effects of the toxicant alone. This could be observed only for phosphamidon, not for the other insecticides.

The isobole for dicrotophos has a sloping symmetry and intersects the phosphamidon curve at middle-to-high toxicant dosages. At the point of intersection both vinyl phosphates are of equal synergisability; both in the high and in the low range of toxicant dosages, dicrotophos is less readily synergised than phosphamidon.

This is to a certain extent in disagreement with Sun's and Johnson's (1960 and 1965) results, but use of another synergist, other techniques, and principles of experimentation can easily account for the differences. A biochemical model for the mode of action of propynyl oxime ethers has not yet been worked out. SACHER *et al.* (1968) suggested a similar mode of action of the propynyl oxynaphthalenes as has been established for the methylenedioxy compounds and C-19795 shows certain structural similarities with these compounds; however, such speculations are lacking all experimental evidence at the moment and can only be regarded as indications for future work.

Summary

Phosphamidon is a water-soluble enolphosphate with systemic action. In this work its specific activity against certain lepidopterous species is discussed, in particular that against the pyralid *Chilo suppressalis* and the olethreutid *Carpocapsa pomonella*. Its effectiveness against these species is in contrast with its specifically small activity against most noctuids, such as *Barathra brassicae*. High toxicity of phosphamidon against homoptera such as *Nephotettix bipunctatus* and *Myzus persicae* is also discussed.

Phosphamidon and dicrotophos are more toxic to *Musca domestica* by oral than by topical application. Conversely, the carbamates dimetilan and dioxacarb are more toxic as contact poisons.

The aphid *Myzus persicae* is able to recognize and avoid phosphamidon in artificial nutrient solutions when probing through a Parafilm membrane.

Parasites and predators are as sensitive to phosphamidon as to other standard insecticides. Bees are not overly sensitive.

The protective mechanism in OP-resistant mites and ticks is effective against all OP-compounds tested including the enolphosphates; monocrotophos showed the greatest activity against resistant forms.

Phosphamidon shows synergism when mixed with synergists in all possible ratios. With sesamex it has a slightly lesser, with C-19795 a better synergisability than dicrotophos.

Résumé *

Effets toxiques du phosphamidon sur les insectes et les mites

Le phosphamidon est un phosphate énolique soluble dans l'eau, à effet systémique. Dans cette étude se trouve une discussion sur son action spécifique contre certains espèces de lépidoptères, en particulier contre la pyralide *Chilo suppressalis* et la olethrentide *Carpocapsa pomonella*. La bonne efficacité contre ces espèces contraste avec l'activité spécialement faible sur la plupart des noctuides tels que *Barathra brassicae*. La toxicité élevée du phosphamidon pour les homéoptères tels que *Nephotettix bipunctatus* et *Myzus persicae* a également été discutée.

Pour *Musca domestica,* le phosphamidon et le dicrotophos sont plus toxiques par voie orale que par application topique. D'autre part, les carbamates dimetilan et dioxacarb sont plus toxiques par contact que par

---

* Traduit par J. P. LANG.

application orale. Le puceron *Myzus persicae* est capable de reconnaître et d'éviter le phosphamidon dans des solutions nutritives artificielles, quand sa trompe pénètre par une membrane de parafilm. Les parasites et les prédateurs sont également sensibles au phosphamidon, comme c'est le cas pour d'autres insecticides standards. Les abeilles ne sont pas spécialement sensibles.

Le mécanisme de protection des mites et des tiques résistantes aux insecticides organophosphorés, fonctionne avec tous les composés phosphorés, y compris les phosphates énoliques. Parmi ceux-ci le monocrotophos présente la plus grande efficacité sur les formes résistantes.

L'effet insecticide du phosphamidon est amplifié quand il est mélangé avec des synergistes appropriés, en toutes proportions. Par rapport au dicrotophos, l'effet du phosphamidon est moins augmenté par le sesamex et plus augmenté par l'éther propinylique du 2,6-dichlorobenzaldoxime (C-19795).

Zusammenfassung [*]

Die toxische Wirkung von Phosphamidon auf Insekten und Milben

Phosphamidon ist ein wasserlösliches Enolphosphat mit systemischer Wirkung. In vorliegender Arbeit wird seine spezifische Wirkung gegen gewisse Lepidopterenarten diskutiert, besonders diejenige gegen die Pyralide *Chilo suppressalis* und die Olethrentide *Carpocapsa pomonella*. Die gute Wirkung gegenüber diesen Arten steht im Gegensatz zur spezifisch geringen Aktivität gegenüber den meisten Noctuiden wie zum Beispiel *Barathra brassicae*. Die starke Toxizität des Phosphamidon gegenüber Homopteren wie *Nephotettix bipunctatus* und *Myzus persicae* werden ebenfalls besprochen.

Phosphamidon und Dicrotophos sind gegenüber *Musca domestica* bei oraler Gabe toxischer als bei topikaler Anwendung. Umgekehrt wirken die Carbamate Dimetilan und Dioxacarb toxischer als Kontaktgifte.

Die Aphide *Myzus persicae* vermag Phosphamidon in künstlichen Nährlösungen zu erkennen und zu meiden, wenn sie durch eine Parafilm-Membrane in jene probeweise einsticht.

Parasiten und Prädatoren sind gegenüber Phosphamidon gleich empfindlich, wie gegen andere Standard-Insektizide. Bienen sind nicht außergewöhnlich empfindlich.

Der Schutzmechanismus in OP-resistenten Milben und Zecken wirkt gegenüber allen geprüften OP-Verbindungen einschließlich der Enolphosphate; Monocrotophos zeigt die größte Wirksamkeit gegenüber resistenten Formen.

Phosphamidon zeigt Synergismus bei Mischungen mit Synergisten in allen möglichen Verhältnissen. Mit Sesamex zeigt es eine etwas geringere, mit C-19795 eine bessere Synergierbarkeit als Dicrotophos.

---

[*] Übersetzt vom Autor.

References

BARTLETT, B. R.: The contact toxicity of some pesticide residues to hymenopterous parasites and coccinellid predators. J. Econ. Entomol. 56, 694 (1963).

— Toxicity of some pesticides to eggs, larvae, and adults of the green lacewing, *Chrysopa carnea.* J. Econ. Entomol. 57, 366 (1964).

— The toxicity of some pesticide residues to adult *Amblyseius hibisci,* with a compilation of the effect of pesticides upon phytoseiid mites. J. Econ. Entomol. 57, 559 (1964).

BERAN, F.: Selektivität einiger Phosphorinsektizide mit besonderer Berücksichtigung ihrer Bienentoxizität. Pflanzenschutzber. 23, 37 (1965).

BROWN, A. W. A.: Insecticide resistance-genetic implications and applications. World Rev. Pest. Control 6, 104 (1967).

DITTRICH, V.: A comparative study of toxicological test methods on a population of the two-spotted spider mite. J. Econ. Entomol. 55, 644 (1962).

ELDEFRAWI, M. E., A. H. HOSNY, A. TOPPAZADA, and S. HASSAN: Susceptibility to acaricides of the mite *Tetranychus cinnabarinus* infesting cotton in Egypt. J. Econ. Entomol. 58, 1106 (1965).

GHOBRIAL, A., V. DITTRICH, M. HAFIZ, H. ATTIAH, and G. VOSS: Population analyses of resistance patterns in spider mites of the *Tetranychus telarius* complex (red and green forms) occuring in Egypt. J. Econ. Entomol. 62, 1262 (1969).

HALL, W. E., and Y.-P. SUN: Mechanism of detoxication and synergism of Bidrin insecticide in houseflies and soil. J. Econ. Entomol. 58, 845 (1965).

HOUGH, W. S.: Toxicity of some insecticides to larvae of codling moth after they enter apples. J. Econ. Entomol. 55, 378 (1962).

JAYCOX, E. R.: Effect on honey bess of nectar from systemic insecticide-treated plants. J. Econ. Entomol. 57, 31 (1964).

KOJIMA, K.: Studies on the selective toxicity of organophosphorus compounds. Special Rept. Inst. Agr. Chem. Toa Noyaku Co, pp. 1–126 (1961).

—, and Y. NAGAE: Reports on the biological effects of phosphamidon, Ciba 885 and other organophosphorus insecticides. Special Rept. Inst. Agr. Chem. Toa Noyaku Co, pp. 1–28 (1958).

KOOY, H. J., JR. (to *Hoffmann-La Roche),* Dutch Pat. No. 6601-926 (Aug. 17, 1966).

LINDGREN, P. D., and R. L. RIDGWAY: Toxicity of five insecticides to several insect predators. J. Econ. Entomol. 60, 1639 (1967).

LOEWE, S., and H. MUISCHNEK: Über Kombinationswirkungen. Arch. exp. Path. Pharmak. 114, 313 (1929).

METCALF, R. L.: Mode of action of insecticide synergists. Ann. Rev. Entomol. 12, 229 (1967).

MITTLER, T. E., and R. H. DADD: Gustatory discrimination between liquids by the aphid *Myzus persicae* (Sulzer). Ent. Exp. & Appl. 7, 315 (1964).

MOOREFIELD, H. H.: Synergism of the carbamate insecticides. Contrib. Boyce Thompson Inst. 19, 501 (1958).

MÜLLER, P.: Der Einfluß von Piperonylbutoxid auf die Wirkung einiger substituierter Phenyl-N-Methyl-Carbamate. Angew. Parasitol. 8, 101 (1967).

PARRY, W. H., and J. B. FORD: The artificial feeding of phosphamidon to *Myzus persicae:* II. The effects of phosphamidon on liquid uptake through a Parafilm membrane. Ent. Exp. & Appl. 12, 1 (1969).

SACHER, R. M., R. L. METCALF, and T. R. FUKUTO: Propynyl naphthyl ethers as selective carbamate synergists. J. Agr. Food Chem. 5, 779 (1968).

SHAW, R. D., M. COOK, and R. E. CARSON: Developments in the resistance status of the southern cattle tick to organophosphorus and carbamate insecticides. J. Econ. Entomol. 61, 1590 (1968).

SUN, Y.-P.: Toxicity index — An improved method of comparing the relative toxicity of insecticides. J. Econ. Entomol. 43, 45 (1950).

—, and E. R. JOHNSON: Synergistic and antagonistic actions of insecticide-synergist combinations and their mode of action. J. Agr. Food Chem. 4, 261 (1960).

SUN, Y.-P., and E. R. JOHNSON: Integration of physico-chemical and biological techniques in specific bioassay, with special reference to Bidrin insecticide. J. Econ. Entomol. **58,** 838 (1965).
— — Relationship between structure of several Azodrin® insecticide homologues and their toxicities to houseflies, tested by injection, infusion, topical application, and spray methods with and without synergists. J. Econ. Entomol. **62,** 1130 (1969).
TAMMES, P. M. L.: Isoboles, a graphic representation of synergism in pesticides. Neth. J. Plant Pathol. **70,** 73 (1964).
VOSS, G.: Weitere Untersuchungen zum Verhalten von Dimecron, Carbicron und Nuvacron in und auf Pflanzen. Unpublished report *CIBA*, Agr. Chem. Div., Basle, Switzerland (1968).
WIESMANN, R.: Neue Mittel und Methoden zur Fliegenbekämpfung im Stall. Schweiz. Arch. Tierheilkde. **102,** 134 (1960).
ZSCHINTZSCH, J.: Der Einfluß von Piperonylbutoxid und anderen Pyrethrum-Synergisten auf die insektizide Wirkung einiger organischer Phosphorsäure-Derivate. Arzneimittelforsch. **11,** 579 (1961).

Chapter 10

Experimental and practical experience with phosphamidon in the field

By

F. BACHMANN

Contents

I. Introduction

Thirteen years have passed since phosphamidon was presented for the first time (BACHMANN 1957). In this period there has been ample opportunity to establish the practical value of the known biological properties of this insecticide and it is certainly not too soon to assess progress and to attempt to show where phosphamidon has proved successful and which were and are the properties that determine its use. The wealth of material available makes it necessary to limit discussion to the most important points and to go into details only where unexpected effects are involved, some of which may not be widely known.

II. Insecticidal properties which determine the range of application

Phosphamidon is a systemic, broad-spectrum insecticide but one which shows especially marked effects against certain species, and against such insects it competes particularly successfully with other insecticides. In this review the emphasis is placed on these main effects.

a) The effect against sucking pests

1. Aphids. — Phosphamidon has made a good name for itself as an aphicide in fruit, field, and vegetable crops; and when it is not quite as effective as some other systemic phosphoric esters against virus vectors, this is usally due to the shorter, by a few days only, duration of effect. Good results that have been obtained are listed in numerous publications, e. g., MEIER (1962) and THURSTON (1965).

The resistance of *Myzus persicae* and *Aphis pomi* to phosphorus esters has recently been increasing in the Mediterranean countries, forcing the substitution of other preparations for phosphamidon.

2. Other homoptera. — Phosphamidon is highly effective against *Psyllidae* and is used regularly for the control of these pests in fruit grow-ing in certain countries. In the Mediterranean countries, however, *Psylla piri* has increasingly developed resistance over the last three years. Phos-phamidon has also become quite important in the control of leafhoppers *(Cicadellidae),* particularly in rice and cotton (e. g., PATHAK *et al.* 1967). Phosphamidon also controls whiteflies *(Aleyrodidae)* effectively but it is not suitable for the control of any of the species of soft and armoured scales *(Coccidae).*

3. Heteroptera. — Though it has a good effect, phosphamidon is sel-dom used for the control of bugs since more persistent contact insecticides, particularly those from the organochlorine group are preferred. ANANTHA-KRISHNAN (1964) discusses its value against this group.

4. Thrips. — The same applies as for heteroptera. Despite phosphami-don's good effect (e. g., DAVIS *et al.* 1966), preference is generally given to DDT for reasons of price and persistance.

5. Spider mites. — Our own experiences, confirmed by those of other experimenters (e. g., FORSYTHE 1965), show that phosphamidon is highly effective against spider mites, particularly *Panonychus ulmi* and *Tetra-nychus urticae,* where the populations concerned have not developed resis-tance to phosphoric esters. But since this is the case in many areas in the world, we must more and more be satisfied with the advantage, which will be discussed later, that the use of phosphamidon against other pests does not encourage an increase in spider mites.

b) The effect against beetles

1. Scarabs. — Although phosphamidon usually has a low contact in-secticidal effect, there are noteworthy exceptions, for example against scarabs. Table I shows the good contact and stomach poison effect and these properties can also be demonstrated in practice (HAGNAUER 1959 a and b).

Table I. *Laboratory experiment with different insecticides as contact and stomach poisons against adults of Melolontha melolontha L.*

Insecticide	Active ingredient (%)	Method [a]	% Mortality after hours					
			2	4	8	24	48	72
Untreated		1	0	0	0	0	0	0
		2	0	0	0	0	0	0
Phosphamidon	0.02	1	20	65	92	100		
		2	25	30	40	60	85	100
Monocrotophos	0.02	1	2	2	12	60	85	100
		2	15	25	37	77	95	100
Lannate	0.02	1	60	65	75	75	97	100
		2	35	62	70	87	97	100
Endosulfan	0.035	1	5	5	5	90	100	
	0.018	1	5	10	10	25	75	90
	0.035	2	0	2	10	60	95	100
	0.018	2	0	2	5	12	45	60

[a] Method 1 = insects are dipped in watery solution and placed on untreated food, method 2 = untreated insects are placed on treated food which is replaced as necessary, food plant = beech, number of test insects = 20 ♂ ♂ + 20 ♀♀.

2. Leaf beetles and flea beetles *(Chrysomelidae).* — Phosphamidon also effectively controls members of this family, with the practical consequence that is has been successfully introduced in Switzerland and in other countries against the potato beetle, particularly against DDT-, lindane- and dieldrin-resistant populations. Table II shows the results of a field trial carried out in Switzerland in 1969 against resistant forms.

Table II. *Field trial against colorado beetle in Basle, Switzerland (1969)* [a]

Insecticide	Active ingredient (%)	Living larvae/plot after 3 days [b]					
		A	B	C	D	⊖	Efficacy (%)
Dioxacarb	0.05	1	0	1	0	0.5	99.5
Phosphamidon	0.03	2	3	4	0	2.2	98
Methidathion	0.03	12	0	4	0	4.0	96
Chlorfenvinphos	0.024	36	1	1	0	9.5	90
Phosalone	0.05	60	0	3	0	15.7	84
Untreated	—	75	161	76	71	95.7	—

[a] Differences between different products were not significant.
[b] Four replicates, plot size 6 × 8 m., treated with 1,500 l./ha. (= 160 gal./acre).

Although it is effective against flea beetles, phosphamidon is not much used in practice due to its shorter persistence compared with the organo-chlorine compounds.

3. Bark beetles. — The control methods against bark beetles have been unsatisfactory up to now since spraying with organochlorine compounds in relatively high doses is undesirable, because of residue problems, and can give rise to disasterous increases in the number of spider mites. All the more reason for mentioning our experience that phosphamidon is most suitable for halting an attack. In April 1968 we noticed that in one of our experi-mental orchards in Switzerland over a dozen prune and plum trees were newly infected with *Anisandrus dispar* F. The beetles had already bored more than one cm. deep and apart from considerable quantities of dust due to the boring, the attack was marked by gummosis from some of the boreholes. Up to 30 of the latter were found/seven-year-old tree, distri-buted about the trunk and main branches. We used this ideal opportunity to carry out an experiment to test different compounds at two different dosages. On May 10, the trunk and main branches were thoroughly sprayed using a knapsack sprayer. No special attention was paid to the foliage. Results showed that complete control could be achieved with 0.1 percent phosphamidon while the other preparations tested, dichlorvos, monocroto-phos, and lindane (0.024 percent) were not so effective. Following this favourable result, less badly affected trees in the same area were treated and, in addition, a number of newly infected seven-year-old apple trees were in other experimental plots. Again, spraying with 0.1 percent phos-phamidon was completely effective.

With this method there is no risk of phytotoxicity or other undesired side effects and it can, therefore, be recommended.

4. Other Coleoptera. — In addition, phosphamidon can be used suc-cessfully against *Blitophaga opaca* L. *(Silphidae)*, *Meligethes aeneus* F. *(Nitidulidae)*, and the alfalfa weevil *(Curculionidae)*.

c) The effect against Lepidoptera

1. Fruit tortrix moths and other Tortricoidea. — Phosphamidon has been used for more than ten years as an insecticide in fruit-growing in Switzerland and recently in increasing quantities in other European coun-tries, particularly in eastern Europe. This increase is due to the fact that phosphamidon is highly effective against codling moth while at the same time controlling sucking pests and certain leaf miners. Table III summarises experiments carried out from 1960 to 1966 in Switzerland and in Yugo-slavia, in which the doses were 0.02 percent active ingredient (a. i.) in Switzerland, where there is only one generation, and 0.03 percent a. i. in Yugoslavia, where there are two generations. Other fruit tortrix moths, for example *Grapholita funebrana* Tr., *Archips rosanus* L. *(Cacoecia ro-sana)*, and *Adoxophyes (Capua) reticulana* Hb., were not sufficiently well controlled with economic doses. Similarly the effect against grape berry

Table III. *Mean efficacy of several insecticides in % against codling moth 1960–1966*

Insecticide	Active ingredient (%)	Efficacy (%)						
		1960	1961	1962	1963	1964	1965	1966
Phosphamidon	0.02–0.03	82	79	78	96	75	92	82
Parathion	0.02	76	—	—	—	—	—	—
Carbaryl	0.075	—	—	97	99	83	—	—
Mobam	0.05	—	—	—	—	—	—	74
Diazinon	0.025	—	—	—	—	—	—	77
% attack in untreated	—		17	12	17	31	15	20	31
No. of trials with phosphamidon	—		2	3	2	4	3	2	2

moth, *Eupoecilia (Clysia) ambiguella* Hb. and *Lobesia (Polychrosis) botrana* Den. and Schiff., particularly against the second generation, is insufficient. *Tortrix viridana* L. and *Choristoneura fumiferana* Clem., the spruce budworm, are very sensitive to phosphamidon and in Canada, phosphamidon has been used for some years to control spruce budworm in areas where the fish toxicity of the insecticide is important (MACDONALD 1965). Phosphamidon can also be used against the bud moths *Spilonota (Tmetocera) ocellana* F. and *Hedya nubiferana* Haw. (*Argyroplace variegana* Hb.), and against *Laspeyresia nigricana* F., the pea tortrix moth.

2. **Leafminers.** — Phosphamidon is highly effective against *Lyonetia clerkella* L., which is the most important member of the *Lyonetiidae* in many fruit-growing areas. Control is achieved with sprays directed mainly against codling moth, as shown in Table IV. However, *Leucoptera cofeella* Guér, coffee leaf miner, which belongs to the same family, is not very sensitive. This may be due to poor penetration of the coffee leaves by phos-

Table IV. *Efficacy of phosphamidon against Lyonetia clerkella L. (apple leaf miner), second generation, Stadel, Switzerland, in 1967* [a]

Insecticide	Active ingredient (%)	Variety	No. of mines				x̄	Efficacy (%)
Phosphamidon	0.02	Berlepsch	1	2	5	3	2.75	96.7
		Goldparmäne	1	4	5	1	2.75	97.1
Fenitrothion	0.04	Berlepsch	8	2	9	4	5.75	93.1
		Goldparmäne	6	9	4	6	6.25	93.5

[a] Dates of treatments: 29 June and 20 July counting on 10 August all the developed mines on 200 leaves/tree.

phamidon. The third economically important member of this family, *Leucoptera scitella* Zell., and also *Stigmella malella* Staint. *(Nepticulidae)* and *Lithocolletis blancardella* F. *(Gracilaridae)* are not adequately controlled by phosphamidon.

3. Ermine moths. — Phosphamidon is generally highly effective at the minimal dose of 0.02 percent active ingredient against the various species of *Hyponomeutidae*. For *Argyresthia conjugella* Zell., the apple fruit moth, this concentration must be increased two or three times.

4. Gelechid moths. — *Gelechiidae* family has several economically important species which can be successfully controlled by phosphamidon, namely, the pink boll worm *Platyedra gossypiella* Saund. (e. g., ALAM 1966), the potato tuberworm *Phthorimaea operculella* Zell. (e. g., SHOREY et al. 1967), and the beet moth, *Scrobipalpa (Phthorimaea) ocellatella* Boyd (e. g., STANKOVIĆ 1964 and ČAMPRAG 1964).

5. Pyralid moths. — By far the most important use of phosphamidon has been in the control of *pyralid* moths. The important stem-borer species which damage rice, maize, and sugar cane belong to this family. Phosphamidon has been used in large-scale control programs in Pakistan and Indonesia against the rice stem-borer. Recently, almost all the spraying has been by ULV-application from aircraft (JOYCE 1968). The insecticidal effect in the field has been recorded by many authors but it is sufficient to refer here to BALUYUT (1964) and BANG and KAE (1964). RANDOLPH et al. (1967) recorded experiments against *Diatraea saccharalis* F. on sorghum where the results showed that phosphamidon was highly effective.

Many positive results have been recorded with phosphamidon in the control of the European corn borer *(Ostrinia nubilalis* Hbn.), particularly in the U.S.A. (e. g., BRAY 1961).

6. Carpenterworm moths (Cossidae). — The value of phosphamidon in controlling the leopard moth *Zeuzera pyrina* L., a pest which is of particular economic importance in fruit-growing areas in Mediterranean lands, was recognised quite early by LISSER (1962). Basing further work on these results, I was able to show experimentally that 0.5 percent phosphamidon injected into boreholes can also give secure kills of the larger caterpillars of the species *Cossus cossus* L. (BACHMANN 1963).

7. Other Lepidoptera. — Phosphamidon also controls caterpillars of the following families: *Geometridae, Lymantriidae, Pieridae,* and *Lycaenidae*. KELLER et al. (1962 a and b) discuss its value in the control of gypsy moth. In Israel, positive results were obtained in practice against *Virachola livia* Klug *(Lycaenidae)*, an important pest of pomegranates.

Unfortunately, phosphamidon has little effect on the caterpillars of the economically very important Noctuidae family and these have to be controlled with other preparations. The only species being recorded as sufficiently sensitive to phosphamidon is *Alabama argillacea* Hbn. (e. g., GIANNOTTI et al. 1965).

d) The effect against Diptera

1. Fruit flies (Trypetidae). — Phosphamidon in common with some other insecticides, for example dimethoate, is effective against fruit fly larvae, and this is of practical importance in crops where it is better tolerated by the plants than other compounds. For example, this is the case against walnut huskfly *Rhagoletis completa* Cresson (NICKEL and WONG 1966). Its use against olive fly *(Dacus oleae* Gmelin) has the advantage that it leaves practically no residues in the fruit or more significantly in the oil.

2. Rust flies (Psilidae). — The single member of economic importance in this family is *Psila rosae* Fabr., the carrot rust fly. The chemical instability and unlimited solubility of phosphamidon in water would seem to make it quite unsuitable for control of this root fly; however, the increasing importance of insecticide residue problems in carrots, which make a large contribution to baby food, gives phosphamidon a real chance in this area. From numerous residue analyses we can show that the residues are already below 0.01 p.p.m. only one week after a treatment by spraying or drenching. The low persistence of the active ingredient means that repeated treatments are needed at intervals of only two weeks; nevertheless, residue-free carrots can still be produced, with effective control. Table V shows the effect of phosphamidon against *P. rosae*.

Table V. *Results of field trials with phosphamidon against carrot rust fly in Switzerland*

Treatment	Method [a]	Active ingredient (%)	No. of replicates	No. of treatments	Weeks interval	Attack (%)	Efficacy (%)	Location
Treated	A	0.05	2	5	2 (first 3)	3.85	95	WEGGIS (1969)
Untreated control	—	—	—	—	—	75.50	—	
Treated	B	0.04	4	5	2	3.90	72	TROISTORRENS (1968)
Untreated control	—	—	—	—	—	13.90	—	

[a] A = drenching with 1/4 l./one m. row length, B = spraying with 2,000 l./ha. (= 214 gal. per acre). First treatment with A and B when highest leaves reached 15 cm.

3. Leaf miner and stem maggot flies (Agromyzidae, Anthomiidae). — Leaf mining fly maggots are well controlled by phosphamidon. GETZIN (1960) discusses the effect against the serpentine leaf miner *Liriomyza munda* Frick, and MEIER (1961) reports on the spinach leaf miner *Pegomyia hyoscyami betae* Curt., a species which frequently causes damage in sugar beet and turnips, particularly in middle and north Europe. Table IV shows our own exeriments. Certain species of *Atherigona* are stem fly pests on

Table VI. *Results of a field trial against spinach leaf miner on sugar beet in Les Barges, Switzerland, 1966* [a]

Treatment	Active ingredient (%)	No. of active mines			Efficacy (%)
		A	B	x̄	
Phosphamidon	0.01	0	0	0	100
Carbaryl	0.025	0	12	6	93.5
Untreated control	—	93	91	92	—

[a] Treatment on 8. June when existing mines were half developed, counting five days later on 25 plants/plot with two replications.

sorghum. Positive control with phosphamidon has been obtained, particularly in Israel.

4. Gall midges (Itonidae). — Apart from the swede midge *(Contarinia nasturtii* Kieffer), none of the gall midges is successfully controlled with economic doses of phosphamidon.

e) The effect against Hymenoptera

1. Diprionidae. — The diprionid and neodiprionid species important in forestry are very susceptible to phosphamidon. Its introduction is particularly useful in forestry areas where there are fishing waters to be considered, for example in Canada (McLeod 1966).

2. Tenthredinidae. — As a result of various official tests and positive practical results, phosphamidon is registered in many countries for the control of important species of sawflies in pomes and stone fruits.

III. Side effects which determine range of application

Today, with such a wide range of highly effective synthetic insecticides available, the positive and negative side effects are often decisive in making the final choice of an insecticide. These additional properties of phosphamidon are so important that they deserve special consideration.

a) Phytotoxicity

When phosphamidon is applied to certain cherry varieties at a concentration of not less than 0.02 percent active ingredient, it is found that after a few days the intercostal areas of the leaves and the leaf margins begin to turn violet. These violet parts become increasingly necrotic eight to ten days after spraying and finally most of the affected leaves drop off. This phenomenon occurs more frequently during the first few weeks after blossom than just before harvest or later, and its intensity varies from year to year. Therefore, phosphamidon should be applied to cherry trees only after preliminary trials with individual branches have shown that it is well tolerated.

A similar type of discoloration and necrosis has also been seen in the Swazi variety of millet; however, this variety is also very susceptible to other systemic organophosphates and is of minor importance as regards acreage. Preliminary tolerance tests are therefore recommended before phosphamidon is applied to millets.

With these two exceptions, phosphamidon is very well tolerated by plants. Of particular advantage is the fact that the active ingredient is soluble in water, so that no emulsifying agents are required for formulation. Sensitive fruit varieties, which during the critical period of the first four to six weeks after blossom frequently suffer from russeting after treatment with other pesticides, may be safely sprayed with phosphamidon. Similarly the common injunction that fungicides and insecticides should not be combined during this period does not apply to phosphamidon/ fungicide mixtures.

b) Physiological effects on plants

Phosphamidon acts as a synergist with fungicides (BACHMANN 1962 and 1968) and in certain cases has a fungicidal effect of its own, although it shows no fungicidal properties *in vitro*. The most plausible hypothesis for this phenomenon is that phosphamidon acts as an antigen which stimulates the plant to form fungicidal antibodies (cf. Kuč 1961). I should like to add a new example from 1969 to those already quoted. Table VII shows the reduction of scab attack following the use of a combination of captan and phosphamidon in comparison to captan and parathion.

Table VII. *Scab attack on fruits of Golden Delicious apples in the experimental orchard in Stadel, Switzerland, 1969* [a]

Treatment	Infestation (%)
Captan + 0.02% phosphamidon	3.7
Captan + 0.02% parathion	15.7
Untreated control	98.8

[a] Addition of insecticide in the first post-blossom treatment and three times (end of June, middle of July, beginning of August) against codling moth. Counting on 200 fruits/tree, five trees/plot.

Some recent examples can also be added to earlier reports on the influence of phosphamidon on the growth and yield of treated plants. Early in 1967 we planted a plot in our experimental vineyard in Rüdlingen, Switzerland, with new vines of a Riesling × Sylvaner cross. Fourteen weekly sprays against *Peronospora* were carried out. Treatment *A* was 0.1 percent folpet, treatment *B* was 0.1 percent folpet + 0.02 percent phosphamidon, and treatment *C* was 0.1 percent folpet + 0.02 percent monocrotophos. The treatments were replicated eight times. At the end of the period of vegetative growth, the main and side branches of all the vines were measured. The results are listed in Table VIII. In analogous experiments carried out earlier with young Blauburgunder vines it was shown that the difference in growth led to an average 25 percent increase in harvest the following year.

Table VIII. *Treatment of vines with insecticide/fungicide mixtures*

Treatment	0.1% Folpet alone	0.1% Folpet + 0.02% phosphamidon	0.1% Folpet + 0.02% monocrotophos
Av. shoot length (cm.) [a]	357	418	456

[a] Differences between folpet alone and folpet + insecticides were found to be significant, whereas differences between folpet/phosphamidon and folpet/monocrotophos mixtures were not significant.

At the 1968 Entomological Congress in Moscow, Balewsky (1968) reported related experiments carried out in Bulgaria. In five-year old Jonathan apple trees, the number of fruits set after June were reduced to 60/tree by picking off the excess. Groups of five trees were sprayed either five times with zineb alone or with the addition of 0.03 percent phosphamidon for the last three times. Trees which had been treated with the combination produced a harvest which was 29 percent greater in weight than that from trees treated with zineb alone, a significant difference. Also, the scab attack was significantly less in the former group.

In a similar experiment in sugar beets, some plots received 200 g. of phosphamidon + 1,250 g. of zineb, some 200 g. of phosphamidon alone, and some were left untreated. The plots, replicated three times, were 150 sq.m. and were treated twice during July at an interval of 19 days. Aphid infestation remained slight and Cercosporiosis did not appear at all so that the differences in yield were due only to the physiological side effects of phosphamidon. Phosphamidon alone produced a yield increase of 59 percent beet and 66 percent sugar; the corresponding figures for phosphamidon and zineb were 61 percent and 71 percent, respectively. These increases were statistically highly significant.

Pivar and Valenčić (1966) reported similar results in Yugoslavia, although there the yield increases were not so great.

It can be stated that in certain crops treatment with phosphamidon alone or more particularly with phosphamidon plus an organic fungicide produces growth and yield increases which in most cases are of significant economic importance.

c) Beneficial insects and predators

The effect of phosphamidon on beneficial insects and predators has been widely studied during the course of pest-control trials. Phosphamidon was found to be relatively harmless to beneficial insects, as was to be expected in view of its very limited action as a contact poison and its property of disappearing rapidly from the surface of the plant into the interior. Thus, CHABOUSSOU (1961) demonstrated that phosphamidon, while active against the wooly aphid *Eriosoma lanigerum* Hausm., is harmless to its parasite *Aphelinus mali* Hald.

As a result of trials for the control of aphid in red clover, JOHANSEN (1960) placed phosphamidon in the leading group of 20 substances in view of its aphicidal action and its harmlessness to bees and to aphid predators and parasites; the clover treated with phosphamidon also gave the highest seed yield.

Details of a braconid *(Hymenoptera)* species and a coccinellid species that are natural enemies of the cabbage aphid have been given by SHOREY (1963), who found that phosphamidon, unlike parathion and diazinon, is relatively specific in its action against aphids and in its harmlessness to beneficial species.

The lack of any ovicidal action on the part of phosphamidon also applies to the eggs of beneficial insects, as was shown by BARTLETT (1964) in the case of *Chrysopa carnea* Steph., the green lacewing. The same author (1963) has also published results that would appear to contradict the findings quoted above. The reason for the unsatisfactory performance of phosphamidon in these trials, which were conducted with five parasitic *Hymenoptera* species and six coccinellid predators, was probably the fact that the various substances were applied to wax paper rather than to living leaves.

From recent literature we can cite LASTER and BRAZELL (1968), who established that, in various spray programmes in cotton, phosphamidon is less dangerous to predators than other insecticides. Finally, HOYT (1969) showed that in an integrated chemical control programme where *Typhlodromus*, the natural enemy of spider mites was concerned, phosphamidon would be used to advantage.

In a concomitant paper, DITTRICH (1970) concludes on the basis of a number of works on laboratory trials with phosphamidon and beneficial insects that in general phosphamidon is no better or worse a performer than other standard insecticides. In the above-quoted field trials, however, the positive aspects seem to point to the fact that phosphamidon indeed has some merits over comparable insecticides.

The potential hazards of phosphamidon to honey bees have been thoroughly investigated, and the verdict given by BERAN and KLIMMER (1969), in agreement with other workers, is that it is toxic to bees. It is, however, one of the least toxic in this group of bee poisons, as has been repeatedly confirmed by other experimental findings and observations, and as can be demonstrated by citing the "hazard index" of some insecticides: phosphamidon 11, aldrin 155, carbaryl 102, DDT 1.5, diazinon 34.3, dichlorvos 750, dimethoate 38 to 51, fenitrothion 56, lindane 35 to 143, malathion 19, monocrotophos 135, and parathion 32.

To close this section I should like to refer to a paper by RAMARAJE et al. (1967) in which the influence of the insecticides phosphamidon, parathion, DDT, malathion, endrin, and lindane on entomogenous fungi was investigated. Phosphamidon proved to be the least harmful.

d) Spider mites

CHABOUSSOU and other authors have established that after the application of certain plant protection products insect or spider mite populations behave differently on treated plants from the way they behave on untreated plants, regardless of the fact that predators and parasites may be eliminated. So, for example, it is known that on plants treated with carbaryl, spider mites have a shorter development cycle, the average number of eggs increases, the sex ratio changes in favour of the females, and the result is a marked increase in spider mites. In practice it is difficult to separate this effect, termed *trophobiosis* by CHABOUSSOU (1966), from the negative influence due to biozonosis and I should hesitate to state categorically that the following examples are due to one cause or the other.

In our experimental orchard in Stadel, Switzerland, long-term experiments with constant spray programs have been running for seven years to compare the effects of six different schedules. The details are shown in

Table IX. *Attack by European red mite in relation to different spray programs on apples (J = Jonathan, G = Glockenapfel)*

Treatment, date, and variety	Av. no. of mites + eggs/leaf					
	Program 1	Program 2	Program 3	Program 4	Program 5	Program 6
Scab fungicide	Folpet	Captan	Captan	Zineb	Delan	(Captan) folpet
Insecticide	Phospha-midon	Phospha-midon	Parathion	Carbaryl	Azinphos-methyl	Ryania + isolan
8/9/1967 J	1.4	1.9	16.1	8.6	1.4	24.9
6/8/1969 J	6.8	11.6	34.5	19.9	22.8	4.2
11/9/1969 J	17.2	54.2	67.2	65.2	74.0	26.6
6/8/1969 G	10.3	15.8	43.5	50.2	61.4	7.6
11/9/1969 G	46.8	58.0	90.7	72.5	105.8	36.6
7/8/1970 J	14.1	12.9	39.2	38.8	51.4	32.2
7/8/1970 G	29.2	48.2	69.6	81.0	54.3	16.6

Table IX. All the programmes included additional wettable sulphur against apple mildew, which was reduced to 0.1 percent in sprays after blossom. Schedule six, which corresponds to the recommendations for integrated control, has been applied since May 1968 against scab using folpet. Previously captan was used but this had to be replaced due to an increasing spider mite problem. Schedule four has been running only since 1967 with the quoted products, and at that same time thiuram was replaced by delan in schedule five. Schedules one-to-three have remained unchanged for seven years. The insecticide addition to the fungicide spray was applied for the first time immediately after blossom (only carbaryl was applied shortly before blossom) and three times later against codling moth. In Table IX are given the results of various spider mite counts and these show the direct influence of the pesticide used on the increase of this pest. We can see that during the whole experimental peroid of seven years it was only with phosphamidon that there was no excessive increase of mites, with its consequent damage. Azinphos methyl was good initially but showed an unfavourable effect after five years. We can see further from the table that captan, in contrast to folpet, encourages spider mite increase. These trial results were better confirmed by many practical experiences; thus, when phosphamidon is used, and fungicides and other insecticides which encourage spider mites are avoided, the spider mite attack seldom exceeds tolerable limits.

e) Tainting

Phosphamidon has been thoroughly studied in this respect and there is ample evidence for the conclusion that it never affects the taste of fruits or vegetables when applied in accordance with instructions. Taint trials have been carried out by CIBA and also by official agencies in Switzerland and the United States (MURPHY et al. 1961).

IV. Usual methods of application

Although phosphamidon is systemic and is rapidly taken up by plant roots, granular or seed-dressing formulations have not proved suitable. This is because its low chemical stability in the soil and loss by washing out due to its solubility mean that it remains available in the soil for only a short time. The practical methods of application are, therefore, limited to spraying, drenching, and dusting.

a) Spraying

The physical properties of phosphamidon render it ideal for high- and low-volume application since it can be mixed in any ratio with water. The addition of a wetting agent is necessary for the high-volume treatment of certain plants that are difficult to wet, such as conifers, brassicas, coffee, or onions, as the wetting power of phosphamidon sprays at normal concentrations is no greater than that of water. The same applies to very hairy plants such as eggplant and soybean. The use of a wetting agent in a low-volume treatment is normally unnecessary, though it has been found to improve the action of the insecticide in conifer forests. Phosphamidon as a liquid active ingredient is also very suitable for ultra-low-volume application from the air. CIBA has had excellent results with this method over very large areas in Pakistan and Indonesia (JOYCE 1968, JOYCE et al. 1968).

Phosphamidon is also suitable as a bait spray against fruit flies in a mixture with hydrolysed protein.

b) Drenching

Drenching may offer certain advantages in the case of potted plants or freshly transplanted seedlings, and as a control method for carrot rust fly, but otherwise it is not recommended for the same reasons that granular formulations are unsuitable.

c) Dusting

The insecticidal effect of dusts containing one-to-three percent phosphamidon is outstanding. Despite this fact, this method has attained no great practical importance because loss through wind drift is often very large, the toxicological risks are greater, and because the storage stability of phosphamidon dusts is very limited.

Summary

Thanks to the good-to-excellent insecticidal properties of phosphamidon against sucking and chewing pests of plants, the preparation can be used successfully in many crops. It has been shown to be particularly effective as an aphicide but it is also highly effective against various beetle species, against codling moth, and against certain leaf-eating caterpillars and leafminers. Its most important application so far has been in the control of stem borers, particularly in rice. The interesting physiological side effects, in the form of yield increases and synergism with fungicides, as well as its relative harmlessness to parasites and predators, are additional reasons for its use in those cases where, as an insecticide, it is no better than competitive products. For other areas of application, which could not be mentioned in this paper, the reader is referred to the "Dimecron" book published by CIBA.

Résumé *

Expériences et essais pratiques avec le phosphamidon, en plein champ

Le phosphamidon peut être utilisé avec succès sur de nombreuses cultures, grâce à ses propriétés insecticides remarquables contre les insectes

---

* Traduit par J. P. LANG.

suceurs et broyeurs. Il s'est révélé particulièrement efficace en tant qu'aphicide et aussi dans la lutte contre diverses espèces de coléoptères, contre les carpocapses des pommes et diverses chenilles dévoreuses et mineuses de feuilles. Mais c'est dans la lutte contre les pyrales du riz qu'il a trouvé son application la plus importante. Le phosphamidon a des effets physiologiques secondaires intéressants: augmentation de certaines récoltes, effets synergiques avec les fongicides. Son rôle est relativement inoffensif vis-à-vis des parasites et prédateurs. Ses propriétés favorisent son choix, même dans des cas où ses effets insecticides ne surpassent pas ceux de produits concurrents. Pour toutes les possibilités d'utilisation du phosphamidon qui n'ont pu être mentionnées dans cette étude, le lecteur pourra se référer au livre intitulé "Dimecron", publié par CIBA Société Anonyme.

Zusammenfassung *

Experimentelle und praktische Erfahrungen mit Phosphamidon im Freiland

Es wird dargestellt, daß dank der guten bis vorzüglichen insektiziden Wirksamkeit von Phosphamidon gegen zahlreiche saugende und fressende Pflanzenschädlinge das Präparat in vielen Anwendungsgebieten im erfolgreichen praktischen Einsatz steht. Bewährt hat es sich vor allem als Aphizid, ferner gegen verschiedene Käferarten, gegen Apfelwickler und manche blattfressenden Raupen und gegen Blattminierer. Die größte Bedeutung aber hat es bis jetzt zur Bekämpfung von Stengelbohrern, insbesondere an Reis, erlangt. Die interessanten physiologischen Nebenwirkungen in Form von Ertragssteigerungen und synergistischer Unterstützung von Fungiziden, ferner seine relative Harmlosigkeit für Parasiten und Prädatoren, geben in zunehmendem Maße den Ausschlag für seine Verwendung auch in jenen Fällen, wo es als Insektizid oft nicht mehr zu leisten vermag als Konkurrenzpräparate. Für alle Einsatzmöglichkeiten, die in dieser Arbeit nicht erwähnt werden konnten, sei auf das von der CIBA herausgegebene Buch „Dimecron" verwiesen.

References

ALAM, M. Z.: Modern insecticides and their uses. *Department of Agriculture*, East Pakistan, Dacca (1966).

ANANTHAKRISHNAN, N. R.: A short note on the use of systemic insecticide for the control of Helopeltis in tea. Union Plant. Assoc. India, Sci. Dept. (Tea Section). Ann. Admin. Rept. 85 (1964).

BACHMANN, F.: Phosphamidon, ein neuer Phosphorsäureester mit systemischer Wirkung. Proc. IVth Internat. Congress Crop Protection, Hamburg 2, 1153 (1957).

— Nebenwirkungen des systemischen Insektizides Phosphamidon auf Ertrag und Pilzbefall einiger Kulturpflanzen. 14th Internat. Symp. Pflanzenschutz, Gent (1962).

— Neue Methoden zur Bekämpfung holzbohrender Raupen. 15th Internat. Symp. Crop Protection, Gent (1963).

---

* Übersetzt vom Autor.

Bachmann, F.: Beitrag zur Kenntnis der Nebenwirkungen einiger Insektizide. 13th Internat. Entomol.-Kongreß, Moskau (1968).

Balewsky, A.: Synergismus und Ertragssteigerungen mit dem Insektizid Dimecron/ Phosphamidon. 13th Internat. Entomol.-Kongreß, Moskau (1968).

Baluyut, E. M.: Performance of various insecticidal sprayings on the rice stem borer control. Rep. Philippines, 12th Ann. Rept. Agr. Prod. Commission for 1963–64, p. 68 (1964).

Bang, Y. H., and B. M. Kae: Timing of insecticides applied as foliar sprays and in irrigation water against Chilo suppressalis in Korea. J. Econ. Entomol. 57, 706 (1964).

Bartlett, B. R.: The contact toxicity of some pesticide residues to hymenopterous parasites and coccinellid predators. J. Econ. Entomol. 56, 694 (1963).

— Toxicity of some pesticides to eggs, larvae, and adults of the green lacewing, Chrysopa carnea. J. Econ. Entomol. 57, 366 (1964).

Beran, F., and O. R. Klimmer: Pflanzenschutzmittel-Kompendium und Richtlinien für die Gebarung mit Pflanzenschutzmitteln. Der Pflanzenarzt 1–94, Wien, Sept. (1969).

Bray, D. F.: European corn borer control in potatoes. J. Econ. Entomol. 54, 782 (1961).

Chaboussou, F.: Action de divers insecticides et notamment de certains produits endothérapiques vis-à-vis d'Aphelinus mali Hald. évoluant à l'intérieur du puceron lanigère du pommier: Eriosoma lanigerum Hausm. Rev. Path. Vég. 40, 17 (1961).

— Nouveaux aspects de la phytiatrie et de la phytopharmacie. Le phénomène de la trophobiose. Proc. FAO Symp. Integrated Pest Control 1, 33 (1966).

Davis, J. W., C. B. Cowan, Jr., W. C. Watkins, Jr., P. D. Lingren, and R. L. Ridgway: Experimental insecticides applied as sprays to control thrips and the cotton fleahopper. J. Econ. Entomol. 59, 980 (1966).

Dittrich, V.: Toxic effects of phosphamidon to insects and mites. Residue Reviews 36, 9 (1971).

Forsythe, H. Y., Jr.: Present status in control of European red mite in Ohio with summer acaricides. J. Econ. Entomol. 58, 811 (1965).

Getzin, L. W.: Selective insecticides for vegetable leaf miner control and parasite survival. J. Econ. Entomol. 53, 872 (1960).

Giannotti, O., A. Orlando, and D. Puzzi: Noções fundamentais sôbre as pragas da lavoura no Estado de Sâo Paulo e como combatê-las. O Biologico 31, 231 (1965).

Hagnauer, W., and J. Michael: Versuche zur Maikäferbekämpfung mit einem systemischen Insektizid. Schweiz. Landw. Monatshefte 37, 65 (1959).

—, A. Meier, and A. Graf: Weitere Erfahrungen zur Maikäferbekämpfung. Schweiz. Landw. Monatshefte 37, 552 (1959).

Hoyt, S. C.: Integrated chemical control of insects and biological control of mites on apple in Washington. J. Econ. Entomol. 62, 74 (1969).

Johansen, C.: Bee poisoning versus clover aphid control in red clover for seed. J. Econ. Entomol. 53, 1012 (1960).

Joyce, R. J. V.: Trials with ultra low volume spraying of Dimecron 100 in East Pakistan. Pest Art. and News Summaries, section A 14, 257 (1968).

—, L. C. Marmol, M. F. J. Brunicardi, and K. Kinvik: Waterless spraying in East Pakistan using the Decca navigation system. Agr. Aviation 10, 118 (1968).

Keller, J. C., E. C. Paszek, A. R. Hastings, and V. A. Johnson: Insecticide tests against gipsy moth larvae. J. Econ. Entomol. 55, 102 (1962 a).

—, V. A. Johnson, R. D. Chisholm, E. C. Paszek, and S. O. Hill: Aerial spray test with several insecticides against gipsy moth larvae. J. Econ. Entomol. 55, 708 (1962 b).

Kuč, J.: The plant fights back. 13th Internat. Symp. Crop. Protection, Gent (1961).

Laster, M. L., and J. R. Brazzel: A comparison of predator populations in cotton under different control programs in Mississippi. J. Econ. Entomol. 61, 714 (1968).

Lisser, A.: The biology of the leopard moth (Zeuzera pyrina) and its control by insecticides, especially Dimecron. CIBA Limited, Basle (1962).

Macdonald, J. R.: Effects of forest spraying on New Brunswick salmon. Canada Department of Fisheries, Fish Culture Development Branch (1965).

MCLEOD, J. M.: Aerial spraying operations against the Swaine jack-pine sawfly in Quebec. *Canada Department of Forestry*, Bi-monthly Prog. Rept. **22** (1966).

MEIER, W.: Die Rübenfliege *(Pegomyia hyoscyami* Panz. Subsp. *betae* Curt.), ein Schädling im Zucker- und Futterrübenbau. Mitt. Schweiz. Landw. **9**, 1 (1961).

— Vergleichende Prüfung von Blattlausbekämpfungsmitteln mit systemischer Wirkung im Zuckerrübenbau. Zucker **15** (4), 84 (1962).

MURPHY, E. F., A. M. BRIANT, M. L. DODDS, I. S. FAGERSON, M. E. KIRKPATRICK, and R. C. WILEY: Effect of insecticides and fungicides on the flavor quality of fruits and vegetables. J. Agr. Food Chem. **9**, 214 (1961).

NICKEL, J. L., and T. T. Y. WONG: Control of the walnut husk fly, *Rhagoletis completa* Cresson, with systemic insecticides. J. Econ. Entomol. **59**, 1079 (1966).

PATHAK, M. D., E. VEA, and V. T. JOHN: Control of insect vectors to prevent virus infection of rice plants. J. Econ. Entomol. **60**, 218 (1967).

PIVAR, G., and L. VALENČIĆ: Uticaj Dimecronai Phaltana na povecanje prinosa korena, lišča i šečera kod šečerne repe u 1965. godine. Agrohemia **11–12**, Beograd (1966).

RAMARAJE, U. N. V., H. C. GOVINDU, and S. K. S. SHIVASHANKARA: The effect of certain insecticides on the entomogenous fungi *Beauveria bassiana* and *Metarrhizium anisopliae.* J. Invert. Pathol. **9**, 398 (1967).

RANDOLPH, N. M., G. L. TEETES, and E. J. BROOK, JR.: Insecticide sprays and granules for control of the sugarcane borer on grain sorghum. J. Econ. Entomol. **60**, 762 (1967).

SHOREY, H. H.: Differential toxicity of insecticides to the cabbage aphid and two associated entomophagous species. J. Econ. Entomol. **56**, 844 (1963).

—, A. S. DEAL, R. L. HALE, and M. J. SNYDER: Control of potato tuberworms with phosphamidon in Southern California. J. Econ. Entomol. **60**, 892 (1967).

STANKOVIĆ, A., and D. ČAMPRAG: New gradation of *Phthorimaea ocellatella* Boyd in Yugoslavia and efficacy of some modern insecticides (in Serbian). Internat. Symp. Sugar Beet Protection Novi Sad, Hemizacija poljoprivrede 445, Beograd (1964).

THURSTON, R.: Effect of insecticides on the green peach aphid, *Myzus persicae* Sulzer, infesting burley tobacco. J. Econ. Entomol. **58**, 1127 (1965).

Appendix

Common, trade, and chemical names of pesticides mentioned in the present volume

Common name	Trade name	Chemical name
aldrin	Octalene	1,2,3,4,10,10-hexachloro-1,4,4a,5,8,8a-hexahydro-1,4-*endo-exo*-5,8-dimethano-naphthalene
amidithion	Thiocron	O,O-dimethyl S-(2-methoxyethyl-carbamoylmethyl) phosphorodithioate
azinphos methyl	Guthion, Gusathion	O,O-dimethyl S-(4-oxo-1,2,3-benzotri-azin-3-(4H)-ylmethyl) phosphorodithioate
captan	Orthocid	N-(trichloromethylthio) cyclohex-4-ene-1,2-dicarboximide
carbaryl	Sevin	1-naphthyl-methylcarbamate
carbophenothion	Trithion	O,O-diethyl S-[(p-chlorophenylthio)-methyl] phosphorodithioate
chlorfenvinphos	Birlane *(Shell)*, Sapecron *(CIBA)*	2-chloro-1-(2,4-dichlorophenyl) vinyl diethyl phosphate
chlorphenamidine	Galecron	N'-(4-chloro-o-tolyl)-N,N-dimethyl-formamidine
chlorthion	Chlorthion	O,O-dimethyl O-(3-chloro-4-nitro-phenyl)-phosphorothioate
coumaphos	Asuntol, Co-Ral	O,O-diethyl O-(3-chloro-4-methyl-2-oxo-2H-1-benzopyran-7-yl) phosphorothioate
DDT	Gesarol	1,1,1-trichloro-2,2-bis(p-chlorophenyl) ethane
demeton	Systox	O,O-diethyl O-(and S)-2-(ethylthio)ethyl phosphorothioates
DEF	—	S,S,S-tributyltrithiophosphate
demeton methyl	Metasystox	O,O-dimethyl O-(and S)-2-(ethylthio)-ethyl phosphorothioates
diazinon	Diazinon, Basudin	O,O-diethyl O-(2-isopropyl-4-methyl-6-pyrimidyl) phosphorothioate
dichlorvos	Vapona *(Shell)*, Nuvan *(CIBA)*	O,O-dimethyl-2,2-dichlorovinyl phosphate
dicrotophos	Bidrin *(Shell)*, Carbicron *(CIBA)*	O,O-dimethyl O-(2-dimethyl-carbamyl-1-methyl) vinyl phosphate
dieldrin	Octalox	1,2,3,4,10,10-hexachloro-6,7-epoxy-1,4,4a,5,6,7,8a-octahydro-1,4-*endo-exo*-5,8-dimethanonaphthalene
dimefox	Pestox IV, Hanane	N,N,N',N'-tetramethyldiamino-phosphoryl fluoride

(continued)

Common name	Trade name	Chemical name
dimethoate	Rogor, Roxion, Perfekthion, Cygon	O,O-dimethyl S-(N-methylcarbamoyl-methyl) phosphorodithioate
dimetilan	Dimetilan	1-(dimethylcarbamoyl)-5-methyl-3-pyrazolyl dimethylcarbamate
dioxacarb	—	2-(1,3-dioxolan-2-yl) phenyl N-methylcarbamate
dioxathion	Delnav, Navadel	S,S'-p-dioxane-2,3-diyl O,O-diethyl phosphorodithioate
disulfoton	Di-Syston	O,O-diethyl S-2-[(ethylthio) ethyl] phosphorodithioate
dithianon	Delan	dicyano-dithiaanthrachinone
endosulfan	Thiodan	6,7,8,9,10,10-hexachloro-1,5,5a-6,9,9a-hexahydro-6,9-methano-2,4,3-benzodioxa-thiepin-3-oxide
endrin	Endrin	1,2,3,4,10,10-hexachloro-6,7-epoxy-1,4,4a,5,6,7,8,8a-octahydro-1,4-*endo-endo*-5,8-dimethanonaphthalene
EPN	—	O-ethyl O-(4-nitrophenyl) phenyl-phosphonothioate
ethion	NIA 1240, Nialate	O,O,O',O'-tetraethyl S,S'-methylene-bisphosphorodithioate
fenitrothion	Folithion, Sumithion	O,O-dimethyl O-(3-methyl-4-nitro-phenyl) phosphorothioate
folpet	Phaltozid, Phaltan	N-trichloromethyl-thiophthalimide
formothion	Anthio	O,O-dimethyl S-(N-methyl-N-formoylcarbamoylmethyl)-phosphoro-dithioate
isolan	Isolan	isopropyl-methylpyrazolyl-dimethyl carbamate
lindane	gamma-BHC	gamma-isomer of 1,2,3,4,5,6-hexachloro-cyclohexane
malathion	Malathion, Cythion	O,O-dimethyl S-(1,2-dicarbethoxyethyl) dithiophosphate
mecarbam	Murphotox	O,O-diethyl S-(N-ethoxycarbonyl-N-methylcarbamoyl-methyl)-phosphorodi-thioate
menazon	Saphos, Saphizon	O,O-dimethyl S-[(4,6-diamino-s-triazin-2-yl)methyl] phosphorodithioate
methidathion	Ultracide, Supracide	O,O-dimethyl-S-(5-methoxy-1,3,4-thiadiazol-2-(3H)-on-3-yl-methyl)-phosphorodithioate
methomyl	Lannate	methyl N-[(methylcarbamoyl)oxy] thioacetimidate
methyl paraoxon	—	O,O-dimethyl O-p-nitrophenyl phosphate
methyl parathion	Nitrox, E-601	O,O-dimethyl O-p-nitrophenyl-phosphorothioate

(continued)

Common name	Trade name	Chemical name
mevinphos	Phosdrin	O,O-dimethyl 1-carbomethoxy-1-propen-2-yl phosphate
mobam	—	4-benzothienyl-N-methyl-carbamate
monocrotophos	Azodrin (Shell), Nuvacron (CIBA)	O,O-dimethyl O-(2-methyl-carbamyl-1-methyl) vinyl phosphate
morphothion	Ekatin F, Ekatin M	O,O-dimethyl S-morpholincarbamoyl-methyl phosphorodithioate
naled	Dibrom	O,O-dimethyl O-(1,2-dibromo-2,2-dichloroethyl) phosphate
oxydemeton methyl	Metasystox R	O,O-dimethyl S-2-(ethylsulfinyl) ethyl phosphorothioate
paraoxon		O,O-diethyl O-p-nitrophenyl phosphate
parathion	Thiophos, Niran, E-605	O,O-diethyl O-p-nitrophenyl phosphorothioate
phencapton	Phencapton	O,O-diethyl S-dichlorophenyl-mercaptomethyl phosphorodithioate
phorate	Thimet	O,O-diethyl S-[(ethylthio) methyl] phosphorodithioate
phosalone	Zolone	O,O-diethyl S-[(6-chloro-2-oxobenz-oxalin-3-yl)-methyl] phosphorodithioate
phosphamidon	Dimecron	O,O-dimethyl O-(1-methyl-2-chloro-2-diethylcarbamyl) vinyl phosphate
piperonyl butoxide	Butacide	butylcarbitol-6-propylpiperonyl ether
piperonyl cyclonene	—	3-isoamyl-5-(methylene dioxyphenyl)-2-cyclohexenone
propyl-isome	—	dipropyl-5,6,7,8-tetrahydro-7-methylnaphtho-(2,3-d)-1,3-dioxole-5,6-dicarboxylate
ronnel	Nankor, Ronnel, Trolene	O,O-dimethyl-O-2,4,5-trichlorophenyl phosphorothioate
ryania	—	Ryania alkaloid mixture
schradan (OMPA)	Pestox III	bis-N,N,N',N'-tetramethyl-diamino-phosphoric acid anhydride
sesamex	Sesoxane	2-(2-ethoxyethoxy) ethyl-3,4-(methylene dioxy) phenyl acetal of acetaldehyde
sulfoxide	Sulfox-Cide	n-octyl-sulfoxide of isosafrole
thiometon	Ekatin	O,O-dimethyl S-2-(ethylthio) ethyl phosphorodithioate
thiram	Thiuram, Arasan	tetramethyl thiuramdisulfide
trichlorfon	Dipterex, Neguvon, Tugon	dimethyl (2,2,2-trichloro-1-hydroxy-ethyl) phosphonate
vamidothion	Kilval, Vation	O,O-dimethyl S-[2-(1-methyl-carbamoyl-ethylthio)-ethyl] phosphorodithioate
zineb	Dithane	zinc-ethylene-bis-dithiocarbamate

Subject Index

Residue Reviews

Previously Published Volumes

Volume 1	0-387-02899-4	$6.10
Volume 2	0-387-03047-6	$6.60
Volume 3	0-387-03048-4	$6.60
Volume 4	0-387-03049-2	$7.20
Volume 5	0-387-03201-0	$8.00
Volume 6	0-387-03202-9	$7.20
Volume 7	0-387-03203-7	$7.20
Volume 8	0-387-03390-4	$8.00
Volume 9	0-387-03391-2	$7.20
Volume 10	0-387-03392-0	$6.60
Volume 11	0-387-03393-9	$8.00
Volume 12	0-387-03647-4	$8.60
Volume 13	0-387-03648-2	$7.20
Volume 14	0-387-03649-0	$8.00
Volume 15	0-387-03650-4	$8.00
Volume 16	0-387-03651-2	$9.10
Volume 17	0-387-03963-5	$10.20
Volume 18	0-387-03964-3	$11.90
Volume 19	0-387-03965-1	$9.90
Volume 20	0-387-04310-1	$11.90
Volume 21	0-387-04311-x	$10.20
Volume 22	0-387-04312-8	$11.60
Volume 23	0-387-04313-6	$12.70
Volume 24	0-387-04314-4	$13.20
Volume 25	0-387-04687-9	$19.80
Volume 26	0-387-04688-7	$10.00
Volume 27	0-387-04689-5	$9.50
Volume 28	0-387-04690-9	$9.50
Volume 29	0-387-04691-7	$13.50
Volume 30	0-387-04692-5	$14.50
Volume 31	0-387-05000-0	$12.00
Volume 32	0-387-05235-6	$14.80
Volume 33	0-387-05236-4	$14.80
Volume 34	0-387-05237-2	$14.80
Volume 35	0-387-05238-0	$14.20
Volume 36	0-387-05373-5	$19.80
Volume 37	0-387-05374-5	$14.80

In Press

Volume 38	0-387-05375-1	$14.20
Volume 39	0-387-05409-x	$13.50
Volume 40	0-387-05410-3	$14.20